KING EDWARD SCHOOL LIBRARY

# DAMS
## BY CASS R. SANDAK

An Easy-Read Modern Wonders Book

FRANKLIN WATTS
New York/London/Toronto/Sydney
1983

*Over:*
Grand Coulee Dam on the Columbia River in Washington.
It is a gravity dam,
and one of the world's largest concrete structures.

FOR MY MOTHER

R.L. 3.9 Spache Revised Formula

Cover photograph courtesy of the Bureau of Reclamation.

Photographs courtesy of: Bureau of Reclamation: pp. 1, 4, 11, 19, 21 (right); Army Corps of Engineers: pp. 9, 24 (bottom left and right); Egyptian Tourist Authority: p. 12; United Nations: pp. 13, 14, 17 (left and right), 21 (left); New York Power Authority: p. 20; New York Public Library: pp. 23, 25 (bottom); Fred J. Maroon: pp. 24 (top), 25 (top); Consulate General of the Netherlands: p. 26 (left and right); British Tourist Authority: p. 27; French Embassy Press and Information Division: p. 28; Manchete/Pictorial Parade: p. 29; Dr. Carl George: pp. 6, 7.

Diagrams by Jane Kendall

Library of Congress Cataloging in Publication Data

Sandak, Cass R.
   Dams.

   (An Easy-read modern wonders book)
   Includes index.
   Summary: Describes various kinds of dams, how they function, and how they are planned and built. Also discusses some of the problems and failures of dams and the uses of other kinds of water barriers.
     1. Dams—Juvenile literature. [1. Dams] I. Title. II. Series.
TC540.S25  1983        627'.8        83-12397
ISBN 0-531-04626-5

Copyright © 1983 by Cass R. Sandak
All rights reserved
Printed in the United States of America
6  5  4  3  2

# Contents

| | |
|---|---|
| What Is a Dam? | 5 |
| Beaver Dams | 6 |
| Parts of a Dam | 8 |
| Types of Dams | 9 |
| Planning a Dam | 12 |
| Building a Dam | 14 |
| The World Inside a Dam | 18 |
| Power from Dams | 20 |
| Dams of the Past | 22 |
| Problems and Failures | 23 |
| Different Kinds of Water Barriers | 26 |
| Dams Today and Tomorrow | 28 |
| Words About Dams | 30 |
| Index | 32 |

# What Is a Dam?

A dam is a wall that holds back water. You can make a kind of dam by cupping your hands around the hole in the sink. Or, when you are in the bathtub, hold your feet together next to the drain. This will stop the water from going down. This is also a kind of dam.

Dams are built across rivers and streams to hold back water. They often form lakes or **reservoirs** that store water. Dams control the supply of water so that there is always the right amount.

People, animals, and plants all need fresh water in order to live. But water is not always easy to find. In some parts of the world, there is very little water. In a desert dams can supply water to make crops grow. Where there is too much water, dams can keep flood waters away from homes and farms.

But most dams do more than just control water. They have many uses and are known as **multipurpose** dams. The water that falls over dams is an important source of electric power. Dams can make shallow rivers deep enough for large ships. Dams also provide areas for fishing, swimming, boating, and other water sports.

Hoover Dam, on the Colorado River, is on the border between Nevada and Arizona. It was once called Boulder Dam. The reservoir behind the dam is Lake Mead.

A beaver dam in the Adirondack Mountains in New York State.

# Beaver Dams

The first dams were built not by people but by beavers. Beavers find a narrow, shallow part of a slow-moving stream. With their sharp teeth they cut down young trees and chew them into sections. They stick the ends in the mud so that the branches point upstream. Beavers then pile up more branches against these upright posts. They fill in the chinks with mud and stones.

Most beaver dams are small, but some are more than 200 or 300 feet (60 to 90 m) long and over 6 feet (1.8 m) high. In the ponds formed by their dams, beavers build dome-shaped lodges out of mud and branches. The dams keep the water deep enough to cover the openings to their homes. This protects beavers from other animals.

Beaver dams are important to wild areas. They can keep soil from being washed away. Beaver ponds fill up when it rains and store water for dry periods. Beaver dams can turn a stream valley into a marsh or swamp. In this way beaver dams change the land.

A beaver dam is never completely watertight. Water seeps through the dam or flows over it. This way the pond stays fresh.

# Parts of a Dam

Sometimes a landslide or earthquake makes a natural dam in a river or stream. But most dams need to be built.

A dam is usually built across a riverbed at a right angle to the way the water flows. The front and back of the dam are called **faces**. The **water face** of the dam is on the **upstream side**. The **downstream side** of the dam is sometimes called the **air face**.

The top of the dam is called the **crest**. Some large dams have roadways built along the crest. This makes a bridge across the river valley.

The sides that slope down from the crest on both sides of a dam are sometimes called the **shoulders**. They give the dam strength.

The **core** is the middle part inside the dam. It is watertight and keeps water from seeping under the dam and washing it away.

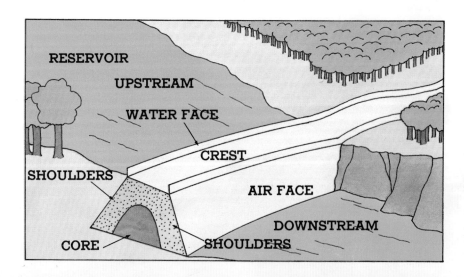

# Types of Dams

Dams are made from earth, rock, or concrete. Concrete is made from powdered stone and pebbles mixed with water and allowed to harden. The concrete is reinforced with metal rods and wires. A few dams have been made from steel or wood, but they do not last long.

An **earth dam** is a giant embankment of soil. Inside it may have a core of packed-down clay or concrete. On the upstream side, earth dams have special watertight layers made from rock. Because earth is lighter than rock and less solid, an earth dam is very large. It is much thicker than it is high.

Fort Peck Dam in Montana was completed in 1940. It is the largest earth dam in the world. It is almost 4 miles (6.4 km) long.

**Gravity dams** use their mass and weight to hold back water. Most gravity dams are made of rock or concrete.

In rocky or mountainous areas, **masonry dams** are built from rock or concrete. Masonry dams span narrow river valleys and are often very high.

Concrete dams are the strongest gravity dams. Like earth dams, they are very big. The thickness at the base is about two-thirds of their height. This means that a 300-foot (91.4-m) high concrete dam would be about 200 feet (61 m) thick at the bottom.

**Rock-fill dams** are built up from loose stone. They have a watertight concrete facing on the upstream side. Under the facing is a layer of stone. The rest of the dam is made up of loose rock.

**Arch dams** use the strength of the arch form. The **crown**, or top, of the arch curves upstream into the reservoir. Water presses against the top of the arch. The stress is transferred to the surrounding rock and to the dam's foundation.

TYPES OF DAMS    (The water face is on the left in each picture)

EARTH DAM

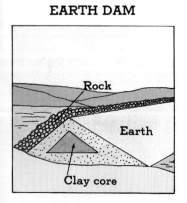

MASONRY DAM

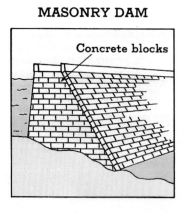

ROCK-FILL DAM

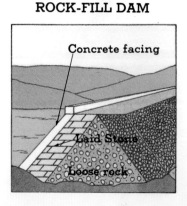

**Single-arch dams** are built where a river or stream passes through a narrow gorge or canyon.

A **buttress dam** is a long dam that has a row of strong supports called **buttresses** on the downstream side. Buttress dams include multiple-arch dams and dams that have an upstream face made up of flat concrete slabs.

Monticello Dam, in California, is a concrete single-arch dam. The arch is laid on its side and curves into the reservoir.

### SINGLE-ARCH DAM
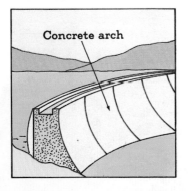

### MULTIPLE-ARCH DAM
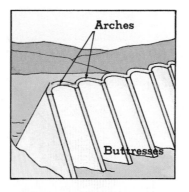

### BUTTRESS DAM
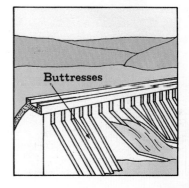

# Planning a Dam

Engineers must pick the location and the type of dam to be built. Surveyors use maps and photographs to study the dam site.

The entire **watershed,** or drainage basin, of the river or stream will be affected. The reservoir site must be cleared. Houses, farms, and businesses need to be moved. Large trees may also be cut down.

When the reservoir formed by the dam is finished, it may cover a very large area—often many square miles.

Lake Nasser, formed by the Aswan High Dam in Egypt, covers more than 2,000 square miles (5,180 sq km).

An irrigation dam under construction in Afghanistan

For safety, engineers design a dam to be even stronger than it needs to be. This added strength helps the dam to stand up in an emergency.

The strength of the rock around the dam is very important. The rock helps hold the dam in place. The weight and pressure of water in the reservoir could rip a dam away from the walls of rock that surround it.

Dams must be able to withstand earthquakes. Scientists have discovered that the weight of water in a dam reservoir pushing the surrounding rock can cause an earthquake.

# Building a Dam

While a dam is being built, small, temporary **cofferdams** help keep the site dry. A wooden or metal framework is filled with earth, gravel, and sand. This framework forms the walls of a large, watertight box built right down to the riverbed.

Cofferdams may also change the course of the flowing water. They force water into **diversion ditches**. These are channels that carry water away from the main streambed where the dam is being built. In rocky and mountainous areas, **diversion tunnels** may have to be blasted through rock to carry water away.

Sometimes a cofferdam looks like a huge barrel sitting in the water on the riverbed. This picture was taken when the Kariba Dam on the Zambezi River in Africa was being built.

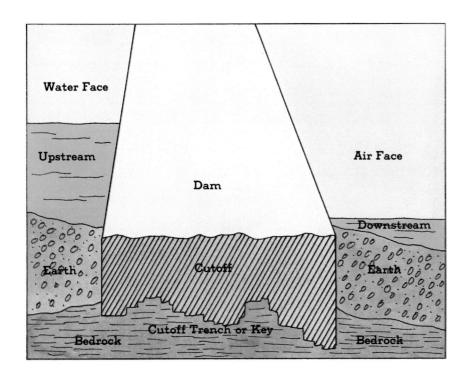

A dam needs a strong foundation of rock or earth that is very firm. Water that seeps under a dam causes **uplift**. This is a force that could loosen a dam from its foundation. Usually workers dig all the way down to the bedrock. Most dams are built in a deep ditch called a **cutoff trench**, or **key**. The trench is cut into the rock with long grooves like those of a key. The grooves keep the foundation from sliding downstream.

The **cutoff** is a watertight foundation built deep into the soil. Sometimes the cutoff extends down 100 feet (30.4 m) into the ground. Liquid cement is forced into holes drilled in the base rock to seal any cracks that might let water through.

An earth dam does not have a cutoff. The core of an earth dam is built up from layers of concrete or clay. The clay used in dams is mixed with sand and water. It is packed down to make it watertight.

Bulldozers and different types of earth-moving equipment build up the upstream and downstream sides of an earth dam. Powerful machines pack down the earth to make it solid and hard. Finally, a layer of stone, called **riprap,** is placed on the upstream side of the dam. The stone must be closely fitted together since this side will eventually be covered with water. The smooth stone facing keeps the parts of the dam made of soil and clay from being worn away by the water.

The part of an earth dam between the upstream and downstream face may be completed either by the **rolled-fill method** or by the **hydraulic-,** or **water-fill, method.** In the rolled-fill method, layers of earth and clay are packed down by machines with heavy rollers.

In the hydraulic-fill method, the running water itself is put to work. A **sluice,** or chute, carries the river water into a trough between the upstream and downstream faces of the dam. The particles of silt, clay, and pebbles in the water settle, dry, and harden. The Fort Peck Dam in Montana was built up in this way.

Two views of the Hoover Dam under construction. The picture at the right shows a diversion tunnel in the lower left corner; the picture at the left illustrates how the dam was built up in blocklike units.

In building a rock-fill dam, rock is heaped up and given a watertight facing. A masonry dam is built up with solid layers of stone. Or else, concrete is poured into wooden frames that build up the dam in blocks. These thin layers harden slowly over a period of days to keep the concrete from cracking.

When a dam is complete, the cofferdams are removed, the diversion ditches are closed off, and the water follows its old course. Only now the dam acts as a barrier. Water collects behind the dam to form a reservoir.

# The World Inside a Dam

A dam is not just a solid wall. There are many pipes and passageways that go right through it. **Penstocks** are large steel pipes that may be 15 feet (4.6 m) in diameter. They are laid directly through the dam as it is built. Water rushing through the penstock pipes turns the **turbines** of **generators**. These generators make electricity.

A mesh filter, or screen, called a **trashrack**, is on the upstream side of the dam. This keeps fish or rubbish from being swept into the openings of the penstocks and clogging them.

Many dams have tunnels, called **scouring galleries**, built into the lower parts. Here the pressure of water flushes away silt that may build up in the reservoir and clog passages through the dam.

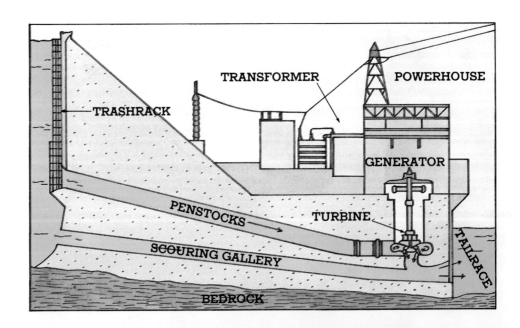

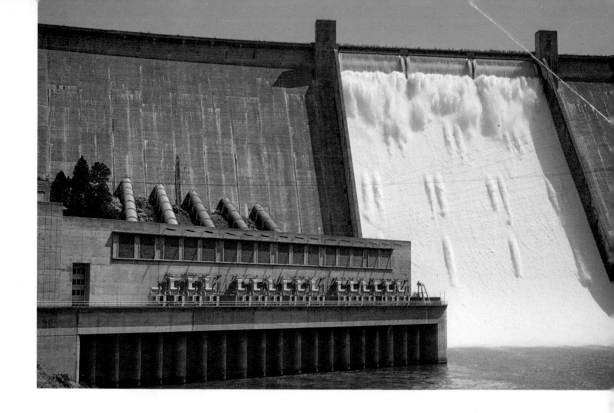

The spillway of the Shasta Dam, a concrete gravity dam on the Sacramento River in California. The powerhouse is at the lower left.

Dam engineers have to remember that heavy rain can swell waters to ten times their normal level. The dam needs to be high enough to stop the water in the reservoir from flowing over its top. **Overtopping** can damage the structure of any dam, but if it is an earth dam, the dam itself can be washed away.

Pipes or troughs called **spillways** keep water from overtopping a dam. They release water from the reservoir before flood levels are reached. The concrete lining of the spillway keeps the dam from being washed away or weakened as water pours through the spillway.

A spillway often has floodgates that can be opened or closed to control the amount of water that passes out of the reservoir.

The Moses-Saunders Power Dam on the St. Lawrence River operates thirty-two generators to supply power to Canada and the United States.

# Power from Dams

Before electricity was discovered, running water supplied mechanical power. Small dams on rivers and streams were fitted with waterwheels attached to millstones for grinding flour or sawmills for cutting timber.

Since the late nineteenth century, dams have been built on most of the world's great rivers to make electricity. These are called **hydroelectric** dams. The word *hydroelectric* comes from the Greek for *water (hydros)* and means electricity made by water. The world's first hydroelectric system was built in Wisconsin in 1882.

Water from the reservoir pours through the penstocks to the powerhouse built near the bottom of the downstream face of the dam. The rushing water spins the blades of the turbine.

The turbine is a type of large waterwheel designed to take advantage of the force and speed of the water. The axle, or shaft, of the turbine turns a coil of wire inside a generator. This is called the **armature** of the generator. It whirls between powerful magnets. This makes electricity.

Transformers send electric current over lines that carry electric power to homes, schools, and businesses that may be many miles away.

Water is not used up by passing over a dam and turning turbines. A pipe or tube under the turbine carries the used water, or **tailwater**, into the **tailrace**, where it flows back into the downstream channel of the river. As long as the water level is high enough to flow through the penstock, more electric power can be produced.

*Left:* inside the Kariba Dam powerhouse. Water will pour through the penstock openings to power a turbine.
*Right:* a giant turbine being lowered into position at the Grand Coulee Dam.

# Dams of the Past

Dams were among the first large building projects undertaken by early civilizations. The ruins of the world's oldest dam are in Egypt along the Nile River near the ancient city of Memphis. They date from about 2700 B.C. and show that the dam was 37 feet (11.3 m) high and 348 feet (106 m) long. Many small dams were probably built in even earlier times, but they were made of clay or dirt that has crumbled and washed away over the years.

Chinese dams date back almost as far as Egypt's. Around 1000 B.C., the Mesopotamians built dams in the Tigris-Euphrates Valley, in what is now Syria and Iraq. The oldest Japanese dam, near Nara, dates from A.D. 162, and is still in use. In India and Pakistan many large dams were built in early times.

The Romans built many dams throughout their empire. After the fall of Rome in A.D. 476, few dams were built for over a thousand years. Then, around 1500, two huge dams were built in Spain. No larger dams were built until the nineteenth century.

A primitive rock dam in China. Wicker nets hold the boulders in place to keep back floodwaters.

Around 1870 a Scottish engineer named William Rankine used mathematics and science in dam building. Before that time dam builders had simply heaped up earth and stone and hoped that their dams would be strong enough. Many new dams were built in the British Isles and in the United States using Rankine's principles.

# Problems and Failures

Dams are built to eliminate problems, but they can also create new problems.

Sometimes dams disrupt life in the communities near the dam site. Over 75,000 people had to leave their homes when Lake Nasser flooded the

The Aswan Dam holds back the rich soil that used to be left behind by the Nile's floodwaters. Egypt's land is less fertile than it once was.

area behind the Aswan High Dam. A project on the Niagara River forced American Indian tribes to move to new areas.

Dams affect animals as well as people. Insects that cause disease can breed in reservoir waters. And dams can destroy wildlife habitats. During Project Noah volunteers in boats saved African wildlife from drowning in the waters that collected behind the Kariba Dam on the Zambezi River.

Since a dam acts as a barrier, it can keep a ship from sailing along its usual route. **Locks** built alongside allow ships to sail around the dam. Locks are short channels with gates at each end.

Dams may disturb historical and archeological sites. Because of the Aswan Dam projects, many ancient Egyptian monuments are now under water.

*Below left:* some kinds of fish swim upstream to spawn. If a dam is in their way, a series of small pools called a fish ladder can help.
*Below right:* locks change the water level to raise and lower ships that pass through.

Lake Nasser flooded many historic sites. The huge Temple at Abu Simbel was moved to higher ground.

And dams can actually cause floods. A dam failure may be a major disaster. When a dam breaks, it suddenly releases a huge amount of water. This can kill people and animals and destroy buildings.

On May 31, 1889, an earth and gravel dam, the South Fork Dam, near Johnstown, Pennsylvania, failed after heavy rains. The rushing water from the 75-foot-high (22.9-m) dam washed away houses and farms. More than 2,000 people lost their lives.

The Johnstown Flood of 1889 was one of the worst disasters in American history.

# Different Kinds of Water Barriers

**Dikes** are long, low special dams that hold back the sea in low-lying areas or in places that are below sea level. Many of these are in the Netherlands.

**Levees** are dikes or embankments that are sometimes built along rivers or streams. Levees keep rivers from flooding their banks, especially when rainfall is heavy and floods threaten.

**Wing dams** are short dams that stretch only partway across a river. They change the speed and direction of water flow. A **weir** is a small dam in a river or stream. It raises the water level or forces the water to flow a different way.

*Left:* a broken dike lets the sea into the low-lying Netherlands. *At right*, a modern flood-control dam protects Dutch farmlands.

The Thames Barrier, finished in 1982, is a completely new type of dam. It is the largest movable water barrier on earth. The barrier protects the city of London, England, from floodwaters that could travel up the river Thames from the North Sea.

The Thames Barrier is made of ten steel floodgates that normally rest at the bottom of the river. They are placed so that the largest ships can sail over them. The 200-foot-wide (61-m) gates stay underwater until steel arms lift the curved floodgates into position.

If the gates of the Thames Barrier were raised into place, they would form a dam more than 50 feet (15.2 m) high.

# Dams Today and Tomorrow

The twentieth century has been the most active period of dam building the world has ever known. Today large dams are still being built all over the world to meet many needs.

Grand Coulee and Hoover dams, two of the world's greatest dams, were built in the United States in the 1930s. About the same time, the French engineer André Coyne invented the thin-arch dam. The Le Gage Dam, built in France in 1954, is 115 feet (35 m) high. At its thinnest point, it is only 4½ feet (1.4 m) thick.

A giant dam project on the Euphrates River in southeastern Turkey will be completed by the year 2000. Two huge dams will triple Turkey's output of electrical power.

A thin-arch dam, the Tolla River Dam in Corsica is 240 feet (73 m) high but is only 8 feet (2.43 m) at its thickest point.

In the future even larger dams will probably be built. Computers can help engineers design these giant dams. Computers can also regulate water levels and flood control.

Soviet scientists have talked about building a dam from Siberia to Alaska across the Bering Strait. This project might be good for shipping and power production. But it could also block ocean currents and affect the earth's climate.

Dams can change the landscape and ways of life for animals and people. And they can destroy. But they also provide electric power, prevent floods, and ensure a water supply. Dams help us to control nature.

The new Itaipu Dam on the Paraná River between Brazil and Paraguay. It will be the world's largest hydroelectric dam when it is operating fully in 1989.

# Words About Dams

**Air face.** The downstream side of a dam.
**Arch dam.** A curved dam that gets its strength from the arch shape.
**Buttress dam.** A dam with a water face of slabs or arches supported by buttresses, or strong supports.
**Cofferdams.** A small temporary dam, like a watertight box or framework, used when dams are being built.
**Core.** The watertight central part of a dam.
**Crest.** The top of a dam.
**Crown.** The part of an arch dam farthest upstream.
**Cutoff.** A watertight dam foundation built deep into the soil.
**Cutoff trench.** A deep trough in which the cutoff is built.
**Diversion dam.** A dam that changes the course of a river.
**Diversion ditch.** A temporary channel that carries water away from a dam site.
**Earth dam.** A large dam made of earth.
**Gravity dam.** A rock or concrete dam that holds back water by its weight.
**Hydroelectric dam.** A dam that makes electricity.
**Hydraulic-fill method (or water-fill method).** A way of building up an earth dam by letting the solid particles in the water settle.
**Key (or cutoff trench).** A trench in which a cutoff wall is built. It has grooves, or notches, like a key.
**Masonry dam.** A dam made of rock or concrete.
**Multiple-arch dam.** A long dam with several arches.
**Multipurpose dam.** A dam that has several uses—water storage, flood control, irrigation, and power generation.
**Overtopping.** Water washing over the top of a dam.
**Penstock.** A channel or pipe that carries water from a reservoir to the turbine of a hydroelectric dam.
**Reservoir.** An artificial lake that stores water behind a dam.

**Riprap.** Closely fitted stone face on the reservoir side of a dam.

**Rock-fill dam.** A dam made of piled rocks with watertight facing of concrete laid over a layer of fitted stone.

**Rolled-fill.** A way of building an earth dam by packing down earth and clay with heavy equipment.

**Scouring gallery.** A tunnel in the bottom part of a dam that flushes silt away.

**Sluice.** A chute that carries water.

**Spillway.** An opening in a dam or reservoir to let out excess water.

**Tailrace.** A channel that carries tailwater back into the main stream.

**Tailwater.** Used water from a hydroelectric dam.

**Thin-arch dam.** A type of thin, high, arched dam made from reinforced concrete.

**Trashrack.** A screen or filter on the upstream side of a dam to keep fish and rubbish out of openings.

**Turbine.** A waterwheel that helps make electric power.

**Uplift.** The upward force caused by water seeping under a dam.

**Water face.** The upstream side of a dam that faces the reservoir.

**Water-fill method.** See hydraulic-fill method.

**Watershed.** The whole area drained by a river or stream. It may cover hundreds of square miles.

# Index

Air face, 8, 30
Arch dams, 10, 30
Armature, 21
Aswan High Dam, 12, 23, 24, 25

Beaver dams, 6–7, 30
Boulder Dam, 4
Buttress dam, 11

Cofferdams, 14, 30
Concrete dams, 9–10
Core of dam, 8, 30
Crest of dam, 8, 30
Crown, 30
Cutoff, 15, 30

Dikes, 26
Diversion dam, 30
Diversion ditches, 14, 30
Diversion tunnels, 14

Earth dam, 9, 30
Electric power from dams, 5

Faces of dam, 8
Floodgates, 19
Flooding from dams, 23–25
Fort Peck Dam, 9, 16

Generators, 18
Grand Coulee Dam, 28
Gravity dams, 9, 30

Hoover Dam, 4, 17, 28
Hydraulic-water-fill, 16, 30
Hydroelectric dams, 20, 29, 30

Irrigation dam, 13

Kariba Dam, 14, 21, 24
Key, 15, 30

Lake Nasser, 12, 25
Levees, 26
Locks, 24

Masonry dams, 10, 30
Multi-arch dam, 30
Multipurpose dams, 5, 30

Overtopping, 19, 30

Penstocks, 18, 30

Reservoirs, 5, 12, 31
Riprap, 16, 31
Rock dam, 23
Rock-fill dam, 10, 17, 31
Rolled-fill, 16, 31

Scouring galleries, 18, 31
Shoulders of dam, 8
Single-arch dams, 11
Sluice, 16, 31
Spillways, 19, 31

Tailrace, 21, 31
Tailwater, 21, 31
Thames Barrier, 27
Thin-arch dam, 28, 31
Trashrack, 18, 31
Turbine, 18, 20–21

Uplift, 15, 31
Upstream side of dam, 8

Water face, 8, 31
Water-fill, 31
Watershed, 12, 31
Weir, 26
Wing dams, 26

D

# THE EARTH AND SPACE

WARWICK PRESS · NEW YORK

*Endpapers: Erosion in the Sahara Desert. Previous page: Sunset over Rio de Janeiro.*

*Above left: The Italian Dolomites. Top: Launch of Apollo 17. Bottom: Hong Kong.*

Published 1979 By Warwick Press Inc. 730 Fifth Avenue, New York 10019
First published in Great Britain by Franklin Watts Ltd. in 1979
Copyright © 1979 by Grisewood & Dempsey Ltd.
Printed in Italy by Vallardi Industrie Grafiche, Milan
All rights reserved 6 5 4 3 2 1
Library of Congress Catalog Card No 78-68539
ISBN 0-531-09144-9
ISBN 0-531-09155-4 lib. bdg.

**Author** David Lambert

**Editorial Consultant** James Muirden

**Editor** Angela Wilkinson

**Illustrators** Tudor Art Agency

# Contents

| | |
|---|---|
| **Chapter One HOW IT ALL BEGAN** | 8 |
| Birth of the Universe | 10 |
| The Layered Earth | 12 |
| The Atmosphere | 14 |
| Oceans | 16 |
| **Chapter Two SHAPING THE SURFACE** | 18 |
| Volcanoes | 20 |
| Earthquakes | 22 |
| Mountains | 24 |
| Land Attacked | 26 |
| Rivers and Ice | 28 |
| **Chapter Three LIFE ON EARTH** | 30 |
| How Life Has Changed | 32 |
| Plants and Climate | 34 |
| Animals of the World | 36 |
| **Chapter Four MAN ON EARTH** | 38 |
| Using the Rocks | 40 |
| Energy | 42 |
| How People Live | 44 |
| The World Laid Waste? | 46 |
| **Chapter Five EXPLORING SPACE** | 48 |
| Rockets | 50 |
| Satellites | 52 |
| Man in Space | 54 |
| **Chapter Six THE MOON** | 56 |
| Exploring the Moon | 58 |
| A Dead World | 60 |
| Moon and Earth | 62 |
| **Chapter Seven THE SOLAR SYSTEM** | 64 |
| Family of Planets | 66 |
| Mercury | 68 |
| Venus | 69 |
| Earth in Space | 70 |
| Mars | 72 |
| The Outer Planets | 74 |
| Junk in Space | 76 |
| **Chapter Eight OUTER SPACE** | 78 |
| Stars | 80 |
| Groups of Stars | 82 |
| Galaxies | 83 |
| Star-Gazing | 84 |
| Space Mysteries | 86 |
| **FACT INDEX** | 88 |

# Introduction

Earth is a very special place. It is the only world we know of with the right ingredients to make life possible. Yet the Earth is just a speck compared with the Sun. The Sun itself is just a small, dimly glowing ball of gas, compared with many other suns.

From the tiny Earth to the giant stars, the universe is full of wonders. **The Earth and Space** guides you through its marvels, from the rocks beneath your feet to clouds of stars far out in space.

Chapter One begins by discovering how our world began. Chapter Two looks at the powerful forces that shaped its lands and seas. Chapter Three briefly looks at life on Earth. Chapter Four shows how people have used the Earth.

Chapter Five tells how people began to explore space. Chapters Six and Seven describe the Moon, Sun and planets. Lastly, in Chapter Eight, we gaze out at the stars beyond the Sun.

Dozens of full-color illustrations help to bring the story vividly alive. Color photographs show such things as the Earth and stars as they actually are. Diagrams help to explain how they were made or how they "work".

Each chapter sets out its story clearly with the help of headings. Under each heading one main idea is explained.

At the end of the book you will find a Fact Index. This will help you to look up items named and shown in the book. But the Fact Index also adds to the information in the main chapters. Turn to this part of the book if you want to check the meaning of words like *igneous*, *orbit* and *quasar*.

# Chapter One
# How It All Began

People often say "as old as the hills" as though the world has never changed. Yet the truth is different. For there was once no Earth, no Sun, and perhaps none of the other worlds and suns now spread through space. No one knows for sure how they all began, but we know how they are changing.

▶ How the universe could have taken shape thousands of millions of years ago. A dense mass may have exploded in a "big bang" (top). From this fireball, huge gas clouds would have spread through space. As the gas clouds cooled they formed into huge groups of stars.

▶ The fiery mouth of a volcano reminds us of the early Earth. At one time molten rock seethed everywhere. Like the stars, the Earth was created from dust and gas spread out through space. Great changes took place before it gained a solid surface.

# Birth of the Universe

In the 1920s, the American astronomer Edwin Hubble discovered that the universe may be getting larger. Distant star groups are speeding off into space, not just away from Earth but away from each other. The matter making up the universe is spreading out, rather like the sides of a balloon when someone blows it up.

Hubble's discovery helped astronomers to work out how the universe might have started. They have two main theories.

**Big Bang or Steady State?**
A Belgian priest called Georges Lemaître suggested that all matter was once squashed together in a mighty fireball. Perhaps 10,000 million years ago the ball exploded,shooting matter out in all directions. As it spread, the matter cooled and came together in the form of the stars. Or so Lemaître believed.

Later, in England, Hermann Bondi, Thomas Gold and Fred Hoyle came up with another theory. Their "steady state" theory suggested that the universe had always existed. As star groups speeded off into space, new ones were born and took their place. Most astronomers accept the big bang theory and think the universe is about 20,000 million years old.

**How the Earth Began**
There are also several theories for the birth of the Earth and the other planets that circle the Sun. One theory says that the Sun and its surrounding planets and moons formed from a cold cloud of gas and dust.

According to this theory, about 5000 million years ago, the force of gravity pulled particles of dust and gas together to produce a huge hot ball. This was the Sun. A flat mass of dust and gas whirled around the early Sun. Dust and gases drew together from eddies in this disk, and took shape as the Earth and other planets. Most scientists accept this explanation.

People once thought that the birth of our solar system was a rare or unique event. But if the disk theory is correct, many other stars must also have systems of planets.

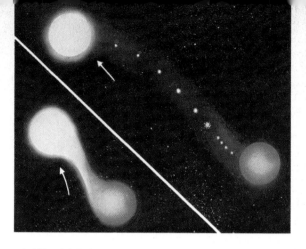

▲ The tidal theory of how the Earth was formed. This theory suggested that a star passed near the Sun, dragging off a tidal cloud of gases. The cloud broke into bits that cooled, condensed, and formed the planets.

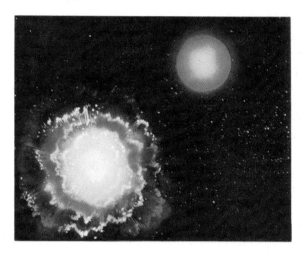

▲ Another theory suggested that the Sun had a companion star. This star exploded and left a cloud of debris that gave rise to the planets, moons, and asteroids. The Sun's gravity stopped them escaping into space.

However the universe began, this fuzzy cloud and five globes show how the Earth probably formed. The first stage was a vast whirling mass of gas and dust. Its particles were drawn in by gravity to form a huge ball. As they became packed closer together, the ball glowed hotly. Later its surface cooled, and dark rafts of solid rock hardened on it. Steam rose from the hot surface, and condensed into clouds that hid the face of the Earth. The atmosphere cooled off, and the clouds shed rain that filled the hollows on the surface of the Earth and formed oceans.

# The Layered Earth

One South African mine is 3840 meters (12,600 ft) deep. No one has gone deeper to see what lies below. But we now know much about the inside of the Earth. Scientists have studied how vibrations travel through the rocks. Some of the vibrations are made by earthquakes. Others are begun by man-made explosions set off underground.

Both kinds of vibration have helped to show that the Earth is layered, something like an onion. There are four main layers—the inner core, outer core, mantle, and crust.

The *crust*, the outer layer, is the ground we stand on. It is a thin outer skin of solid rock covered with soil. The crust under the land is up to four times thicker than that on the floor of the oceans. The crust of the land is mainly made of granite. But the crust beneath the oceans is of a heavier rock called basalt.

The rather light rocks of the crust float like scum upon the hotter, heavier rocks underneath, which make up the *mantle*. No one has seen these rocks. Geophysicists believe that they are solid. But parts of the mantle seem to behave more like soft taffy than hard rock.

Below the mantle lies the layer called the *outer core*. In 1936 scientists discovered that this must be made of very hot, molten material. Most of it is iron and nickel. The rest seems to be a substance called silicon. Below the liquid outer core lies the *inner core*. This is probably completely metallic.

## How the Earth Got its Layers

The layers of the Earth are about 4600 million years old. At first, dust and gas may have built a solid ball. But great pressures crushed the rocks inside and made them very hot. Iron and other heavy substances would then have sunk down to form the core. Lighter substances such as oxygen, hydrogen and silicon would have floated upward towards the surface.

There, different elements combined differently. Some created silicates. These cooled and hardened to create solid rafts of rock that floated on an ocean of denser molten rock. Some of the oldest rocks that we know hardened about 3800 million years ago.

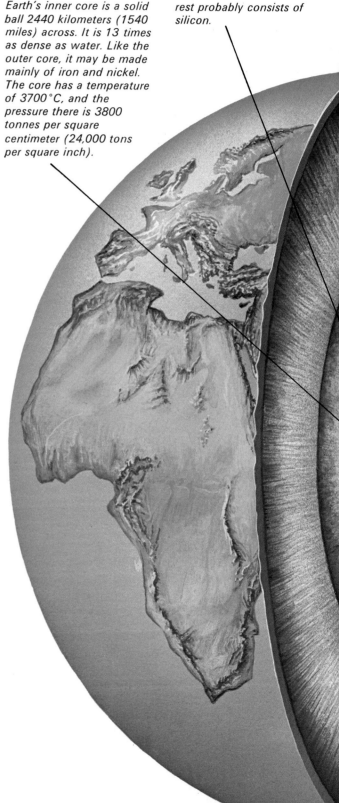

*Earth's inner core is a solid ball 2440 kilometers (1540 miles) across. It is 13 times as dense as water. Like the outer core, it may be made mainly of iron and nickel. The core has a temperature of 3700°C, and the pressure there is 3800 tonnes per square centimeter (24,000 tons per square inch).*

*The outer core lies below the mantle and above the inner core. It is 2240 kilometers (1400 miles) thick. Experts believe that it is mainly made of molten metals which are very hot and under enormous pressure. Four-fifths of it may be iron and nickel. The rest probably consists of silicon.*

The mantle lies beneath the crust and above the outer core. It is nearly 2900 kilometers (1800 miles) thick. It consists of dense, hot rocks. The temperature and pressure here are lower than those of the core. But much of the mantle rock is semi-molten and flows in sluggish currents.

▶ Some of the world's oldest rocks, photographed in Greenland. Such rocks were formed more than 3700 million years ago. They consist of gneiss, a type of rock created from other rocks by heat and pressure.

The crust is the Earth's solid outer layer. It is up to 30 kilometers (19 miles) thick beneath mountains but only 6 kilometers (4 miles) thick under the oceans. Its rocks "float" on the denser rocks that make up the mantle.

### The Cooling Crust

As the Earth's surface cooled, the rock rafts grew larger. At last a hard crust encased the whole globe. But molten rocks still sometimes pushed up to the surface. They spilled out from the holes and cracks now called volcanoes.

Meanwhile, gases—lighter than the lightest rocks—escaped from the inside of the Earth. Above its surface, steam cooled to water vapor and fell back as rain. So began the seas and rivers. The pull of gravity held most other gases close to the Earth's surface. They formed the Earth's early atmosphere.

▼ A slice through the Earth's crust. Two thin layers make up the ocean floors. Beneath the continents the crust is thicker and more complicated.

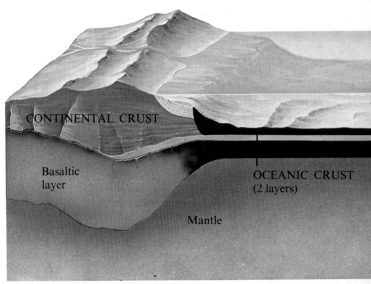

# The Atmosphere

▼ The air we breathe clings to the Earth's surface like a thin, invisible skin. Its main ingredients are nitrogen and oxygen.

Gaze up at a clear night sky. Nothing seems to lie between you and the stars. But in fact you are gazing up through air—the layer of invisible gases making up Earth's atmosphere. Air is chiefly made of nitrogen and oxygen, but there are other gases too. Air also holds tiny particles of dust and water. Balloons carrying instruments have shown scientists that the Earth's atmosphere has four main layers.

## Air for Life and Weather

The lowest layer of the atmosphere is called the *troposphere*. This layer is between 8 and 16 kilometers (5 and 10 miles) thick. Most of the air is concentrated in this layer. Only the troposphere holds enough air for man and other living things to breathe.

Here, too, is the world's weather factory, where the winds and rain begin. The Sun warms up the air more over some parts of the Earth's surface than over other parts. Warmed air rises, and heavier, cooler air flows in to take its place. This causes winds. Warming and cooling also produce water vapor, clouds, rain, sleet and snow (see page 17). The kind of weather a region gets over many years is called its *climate*. The kind of climate a region has influences the kind of plants that grow in it.

## Where Air Grows Thin

Above the cold upper level of the troposphere lies the atmospheric layer called the *stratosphere*. The stratosphere rises to about 64 kilometers (40 miles) above the Earth's surface. It holds a layer of ozone —a form of oxygen that shields the Earth's living things from harmful ultraviolet radiation sent out by the Sun. Above the stratosphere lies a level sometimes called the *mesosphere*. This is the coldest region of the atmosphere.

Above the mesosphere come the even sparser gases of the *ionosphere*. Here, the Sun's fierce radiation ionizes the gas atoms. That is, it knocks off electrons, and makes the atoms electrically charged.

The *exosphere*—the outer layer of the atmosphere—is made up of the lightest gases: helium and hydrogen. Many hundreds of kilometers out, the atmosphere thins to nothing.

**The Air We Breathe**

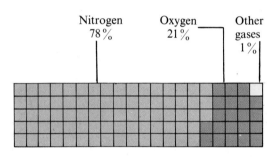

▼ The atmosphere affects the Sun's heat reaching different parts of the Earth. The less air the Sun's rays have to penetrate, the warmer is the place below. Also the more directly the Sun falls on the land, the warmer it becomes. The rays must pass through more air and are more thinly spread above the poles than above the equator. Thus high latitude regions near the poles stay cold. Low latitude regions near the equator stay hot.

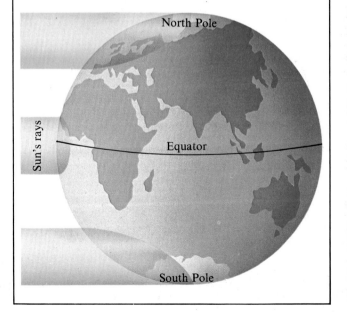

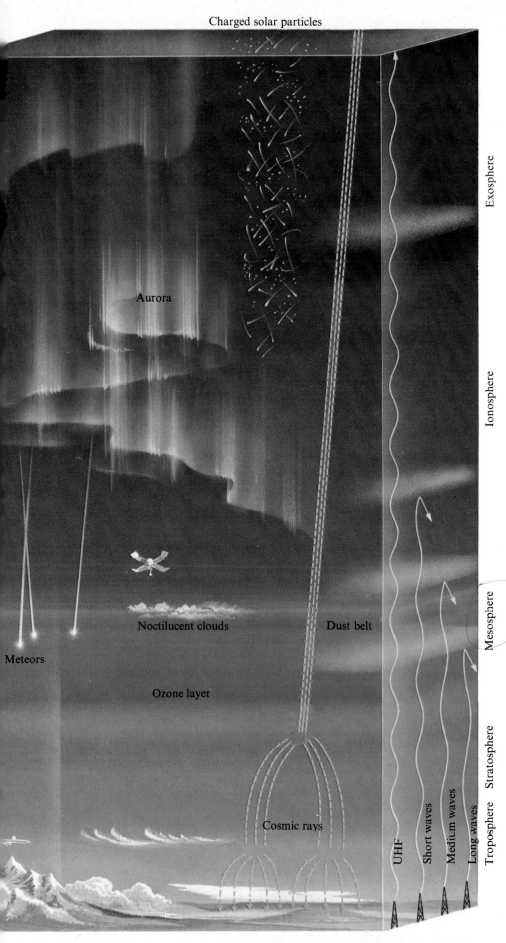

The exosphere is the outer layer of atmosphere above 500 kilometers. Very short-wave (ultra-high frequency or UHF) radio waves pass through it into space.

The ionosphere is between 80 and 500 kilometers (50 and 300 miles) above the ground. Its temperature rises from −80°C at 80 kilometers to 1200°C at 500 kilometers. Inside the ionosphere, cosmic rays and solar rays hit atoms of gas and so produce electrically charged ions. Particles from the Sun which disturb the ionosphere produce the glowing curtains of light called auroras. Dust swirls show up as so-called noctilucent clouds. Air resistance in the ionosphere makes meteors begin to burn. Artificial satellites orbit the Earth at this level, and short, medium and long-wave radio waves are bounced back to Earth from it.

The stratosphere is over 60 kilometers (37 miles) thick. Its temperature changes with height. It is −55°C low down, 10°C at 50 kilometers, and −80°C at 80 kilometers. The warm layer contains ozone, heated by the Sun. This gas absorbs much of the Sun's ultraviolet light. Cosmic rays hit atoms in the stratosphere, creating weaker cosmic rays that bombard the Earth. Cirrus clouds of ice crystals reach 12,000 meters. Otherwise the weather is clear. Jet airliners fly in the lower stratosphere.

The troposhere is 16 kilometers (10 miles) thick at the equator, but only 8 kilometers (5 miles) thick at the poles. Nine-tenths of all air is in the troposphere. At sea level air pressure equals one kilogramme resting on each square centimeter, or about 14 lbs per sq inch. At 18,000 meters (10 miles) up, pressure is only one-tenth as great. As you rise through the troposphere, temperature drops by 1°C for each kilometer.

15

# Oceans

◄ *Two views of the world as seen from space. They help us to realize that the oceans cover nearly three-quarters of the Earth's surface.*

**Ocean Facts and Figures**
Extent of oceans: 71% of Earth's surface
Mean depth: 3554 meters (11,660 ft)
Greatest depth: 11,033 meters (36,197 ft)
Volume: 308,544,000 cubic miles
Weight: 0·022% of the Earth
Largest ocean: Pacific Ocean (63,770,000 sq miles)
Smallest ocean: Arctic Ocean (5,426,000 sq miles)

▼ *A cutaway view of the floor and currents of the North Atlantic Ocean. The great underwater Mid-Atlantic Ridge runs down the middle. Here and there its tips peep from the sea as islands. On each side of the ridge, the ocean floor falls away to form vast abyssal plains. Near the land, the floor rises in a steep slope, up to the shallow continental shelf around North America and Europe. The arrows show ocean currents. The Gulf Stream sends warm water from America to Europe. The blue arrows show the Labrador Current which brings cold water down from the Arctic to chill the northeastern shores of North America.*

Nearly three-quarters of the Earth's surface lies under a great sheet of salt water. Its four main basins make up the oceans. Largest of all is the Pacific. You could fit all the world's lands into its basin. The Pacific is also the deepest ocean. In parts it would be deep enough to drown Mount Everest, the world's highest mountain. Asia and Australasia form the western edge of the Pacific. North and South America are its eastern rim.

The Atlantic Ocean is the second largest ocean. It separates the Americas from Europe and Africa.

The Indian Ocean is the third largest ocean. Africa lies to its west, Asia to its north and northeast, and Australia to its east.

The Arctic Ocean is smaller, shallower and colder than the other three. Much of it lies under ice. Almost all of it is hemmed in by the northern coasts of Asia and North America.

Some geographers give the name Antarctic Ocean or Southern Ocean to the southern ends of the Pacific, Atlantic and Indian oceans.

The oceans hold almost all the world's water. Seawater tastes salty. This is because rivers dissolve salts from the land and wash them into the sea. When seawater evaporates (turns into water vapor) the salts stay in the sea. Thus over millions of years the oceans have become too salty to drink.

The amount of water in the oceans stays more or less the same. The water cycle ensures that water which evaporates under the Sun's heat eventually returns to the oceans.

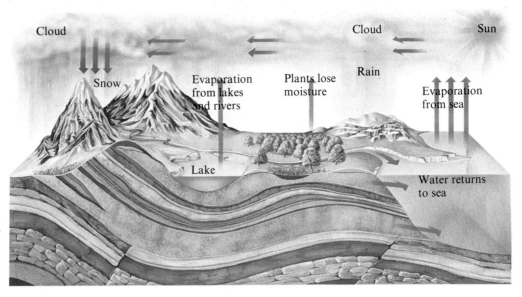

▶ The oceans lose water, but regain it. This process is called the water cycle. It starts when the Sun's heat draws up water vapor from the sea and land. The rising vapor cools and condenses into droplets that form clouds. Clouds may shed their moisture as rain, hail, sleet or snow. The water runs into rivers or seeps into the ground. In time, underground and surface water find their way down to the sea. The water cycle keeps the oceans full.

## Waves and Currents

Sometimes the ocean surface is calm. At other times the wind whips up waves. The waves move through the water. But the water stays where it is as each wave passes by. This movement makes boats bob up and down. Tides act like huge waves in slow motion (see page 63).

Strong, steady winds keep parts of the ocean surface moving as currents. Some carry warm water to cold parts of the world. Others bring cold water to warm areas.

Because the Earth spins from west to east, ocean currents bend to the *right* north of the equator. South of the equator currents bend to the *left*. Some currents flow in huge circles. Cold currents flow deep down, often in different directions from those on the surface. In places, cold water wells up from the depths.

## The Ocean Floor

The sea floor is shallow around the edges of the continents. It forms a great shelf, which, in some places, reaches hundreds of kilometers out into the ocean.

Then comes a sudden drop—a vast cliff where the sea bed falls away to the true ocean floor. This is about 3·5 kilometers (over 2 miles) deep. In places, the floor is torn by even deeper gashes. Deepest of all is the Marianas Trench in the Western Pacific. Here the ocean is 11,000 meters (36,000 ft) deep.

Elsewhere, huge mountain chains rise from the depths. The Mid-Atlantic Ridge is a huge mountain chain that runs from north to south down the middle of the Atlantic Ocean. Here and there a few of its peaks peep out of the sea to form islands. Iceland and the Azores are examples.

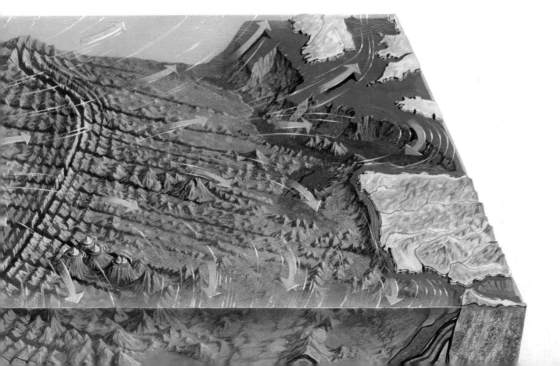

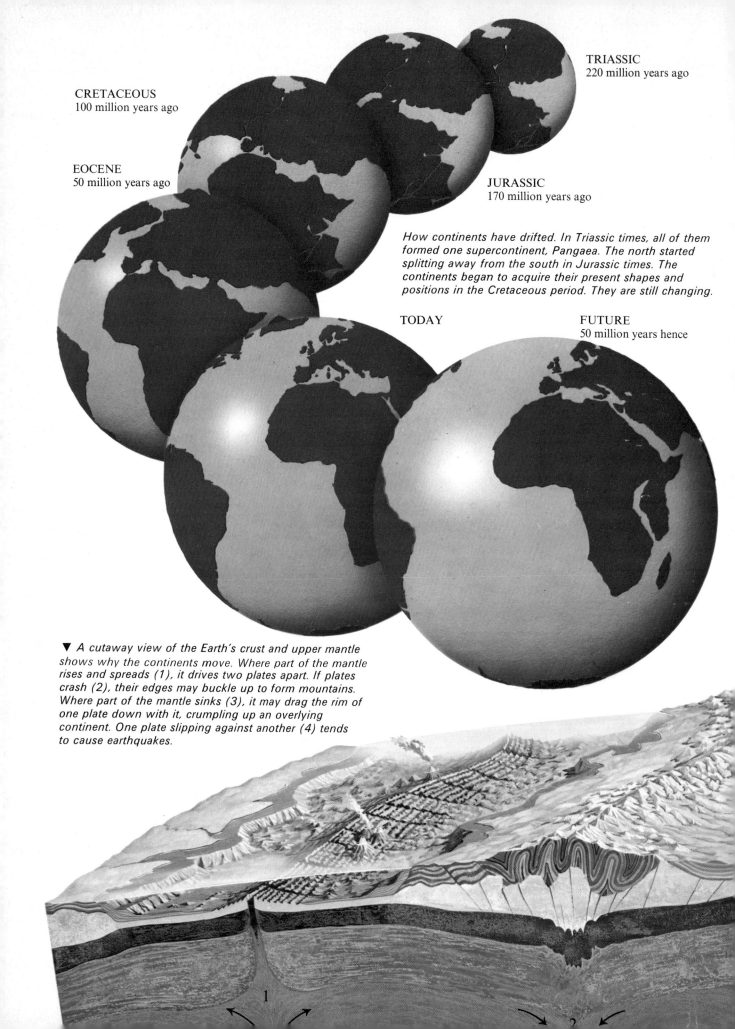

CRETACEOUS
100 million years ago

TRIASSIC
220 million years ago

EOCENE
50 million years ago

JURASSIC
170 million years ago

*How continents have drifted. In Triassic times, all of them formed one supercontinent, Pangaea. The north started splitting away from the south in Jurassic times. The continents began to acquire their present shapes and positions in the Cretaceous period. They are still changing.*

TODAY

FUTURE
50 million years hence

▼ *A cutaway view of the Earth's crust and upper mantle shows why the continents move. Where part of the mantle rises and spreads (1), it drives two plates apart. If plates crash (2), their edges may buckle up to form mountains. Where part of the mantle sinks (3), it may drag the rim of one plate down with it, crumpling up an overlying continent. One plate slipping against another (4) tends to cause earthquakes.*

# Chapter Two
# Shaping the Surface

▼ *The Sinai Peninsula and the Gulf of Suez, as seen by a satellite orbiting the Earth. The Gulf of Suez is a gap now slowly opening to separate Africa from Asia.*

Over many millions of years, the land you stand on may creep half way around the world. Meanwhile, it may be thrust up, worn down, and pushed up once again. Immensely powerful forces have shaped the surface of the Earth.

On a map, the continents look as if they could be fitted together like pieces of a giant jigsaw puzzle. In 1910 a German scientist, Alfred Wegener, suggested a simple explanation. Wegener said that all the continents had once been one great mass of land. He called it Pangaea ("All Earth"). He believed that Pangaea split up into separate continents, which drifted to where they lie now.

Scientists think they know how this *continental drift* happened. Deep down in the Earth, the great heat of its core warms the mantle and sends currents rising upward to the crust. The crust is made up of huge interlocking slabs, called *plates*. The continents and ocean floors are embedded in them. Where a rising current in the mantle reaches the crust, it forces two plates apart.

In this way, the Americas were pushed away from Africa and Europe. Molten rock, called *magma*, welled up to fill the gap that opened in the middle of the ocean floor. The mantle current spread out, cooled, and sank, dragging with it rocks from other regions of the crust. The continents are still traveling, and crust is still being made and lost.

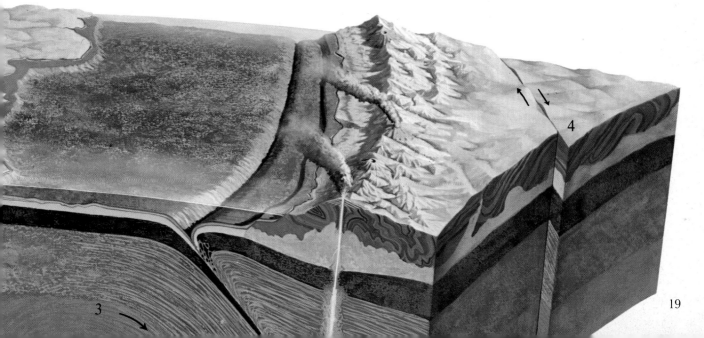

# Volcanoes

Volcanoes are holes in the ground through which hot, liquid rock and gases escape from below the surface. Much of the Earth's crust and many of its mountains originally came from volcanoes.

Volcanoes are fiery clues to the great heat and pressures deep down in the Earth. They occur where the Earth's crust is weakest, especially where two plates meet or separate. Here, pressure may force melted rock and other substances up from the mantle and onto the surface of the Earth.

**Three Kinds of Volcano**
Most known volcanoes stand on land, although some erupt below the sea. Scientists have found that land volcanoes are of three main kinds. One type erupts with sudden, frightening explosions. This is because its molten rock has many hot gases trapped in it.

As the molten rock, or magma, bursts out onto the surface, the gases suddenly expand. They shoot out fragments, like shrapnel from a high explosive shell. These fragments include ash, cinders, and volcanic bombs made of hot lumps of molten rock. Around the hole from which it comes, the ash from an explosive volcano builds up a steep-sided cone.

Other volcanoes are much quieter. Some release little gas. In others, the gas escapes easily. Quiet volcanoes do not produce much ash or many cinders. Nor do they explode violently. Instead, their molten rock, or lava, spills out over the ground from a hole or crack. Runny types of lava may flow a long way before cooling and hardening. The result is a volcano with gently sloping sides. In parts of India and North America, huge old lava flows have spread out smoothly over hundreds of kilometers.

If the lava is more sticky, it quickly hardens. Sticky lava builds a spine or a steep dome.

As well as explosive and quiet volcanoes, there are intermediate volcanoes. They often erupt with an explosion. But they also pour out flows of lava. Their cones often build up like sloping sandwiches made up of ash and lava layers. Mount Vesuvius in Italy and the island of Vulcano off Sicily are two famous volcanoes of this kind.

## Sleeping Volcanoes

In time, volcanoes stop erupting. If they stop for good, scientists describe them as *extinct*. There are extinct volcanoes in most parts of the world.

But certain volcanoes are only *dormant*. They may lie quiet for hundreds of years, and then erupt with sudden violence. A volcano will stop erupting if a plug of solid lava blocks its vent, or outlet. But gas pressure may build up below. Suddenly the gases may blast the top off the volcano. Krakatoa in Indonesia and Thíra in the Aegean Sea are volcanoes that exploded in this way. The force of their explosions was far greater than that of any hydrogen bomb that man has ever made.

> **Volcanoes—Facts and Figures**
> There are 455 active volcanoes on land, and 80 more under the sea.
> The highest volcano is Aconcagua in Argentina (6960 meters, 21,000 feet).
> The biggest crater is Mount Aso, Japan. It is 27 kilometers (17 miles) across.
> The greatest volcanic explosion happened about 1470 BC when the Aegean island of Thíra blew up with the force of 300 hydrogen bombs.

*A cutaway view of one volcano. Pressure forces molten rock (magma) from the mantle to the surface through a weakness in the crust. The magma rises through basalt, granitic, metamorphic and folded rocks. It escapes from a vent in the surface. The volcano mouth may throw out lava, dust, ash, gas, and chunks of solid rock.*

▲ *A molten lava fountain spurts from Mount Kilauea in Hawaii. Beyond, molten lava runs downhill. Kilauea is on the side of Mauna Loa, a huge volcano.*

*The world's volcanoes. Most occur as magma wells up to the surface where plates collide or separate.*

Direction plate is moving
Collision zone
▲ Volcanoes

### DIFFERENT KINDS OF VOLCANO

*Runny lava flows far and builds a shield volcano.*

*Thicker lava and ash form layers that build up a cone-shaped volcano.*

*Very thick lava may harden and plug the volcano until pressure blasts it out.*

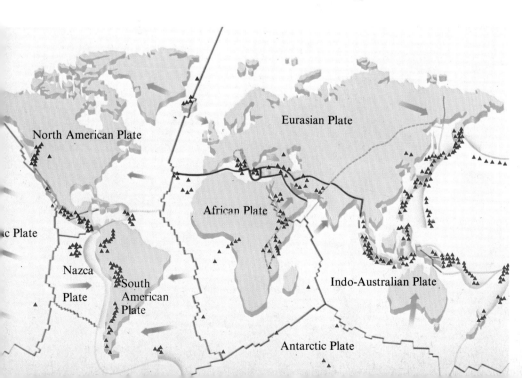

# Earthquakes

"Safe as houses" is a saying only true in certain countries. From Mediterranean lands to South-East Asia and around the rim of the Pacific Ocean, houses sometimes tumble as the ground they stand on heaves and splits apart. In some years, earthquakes somewhere in the world destroy thousands of lives and buildings. But most earthquakes are too slight to be felt.

Like volcanoes, earthquakes tend to happen where the great plates of the Earth's crust slowly crash together. Where plate rims are dragged below the ocean trenches, great forces set up strains in the surface rocks. At times, these forces tear a gash across the land or sea bed. At other times, the rock on one side of such a crack or *fault* moves slightly but suddenly up, down, or sideways.

Such movements produce a shock that travels through the crust and sets the surface trembling. Most earthquakes have their origin or focus within 20 kilometers (12 miles) of the surface. Some are even launched by a landslide or an avalanche. Others may be set off by some disturbance deep inside Earth's mantle.

The most damaging earthquakes come from less than 60 kilometres (37 miles) down. A sea-bed earthquake may set off a giant ocean wave called a *tsunami*. Tsunamis can be much higher than a house and travel faster than the fastest train. Tsunamis may drown thousands of people.

**Earthquakes—Facts and Figures**
About half a million earthquakes are recorded every year. One in five can be felt. Only one in 500 does any damage.
The severest earthquake in modern times struck Alaska in 1964. It measured 8·9 on the Richter scale and unleashed the mightiest known ocean wave—67 metres high.
The most earthquake deaths have occurred in China. In 1556 about 800,000 died in Shensi Province. In 1976 nearly as many may have died at Tangshan.

▼ *Three tracking stations (dots) detect the distance of an earthquake (circles) by measuring the waves it sends through the ground. Its epicenter is at the surface where the circles meet, above its focus.*

▼ *A seismograph shows earth tremors as wriggles in a line traced on a turning drum. A tremor vibrates the drum but not the weight that holds the tracer.*

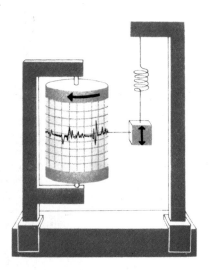
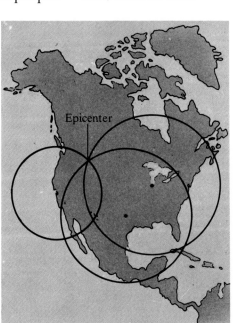

◄ *Buildings shattered and tilted by the San Francisco earthquake of 1906. This tremor and the fires that followed killed 700 people and destroyed 497 buildings.*

▼ *In California, two plates meet at the San Andreas Fault. The western plate moves northwest, but is forced west and then north where it meets the Sierra Nevada Mountains. The movements of this plate against its neighbor have caused thousands of earthquakes.*

▶ *On this small-scale map of California, thick lines show the main lines of weakness responsible for California's earthquakes. Movements of land on both sides of such lines helped to produce the blocks of land shown below.*

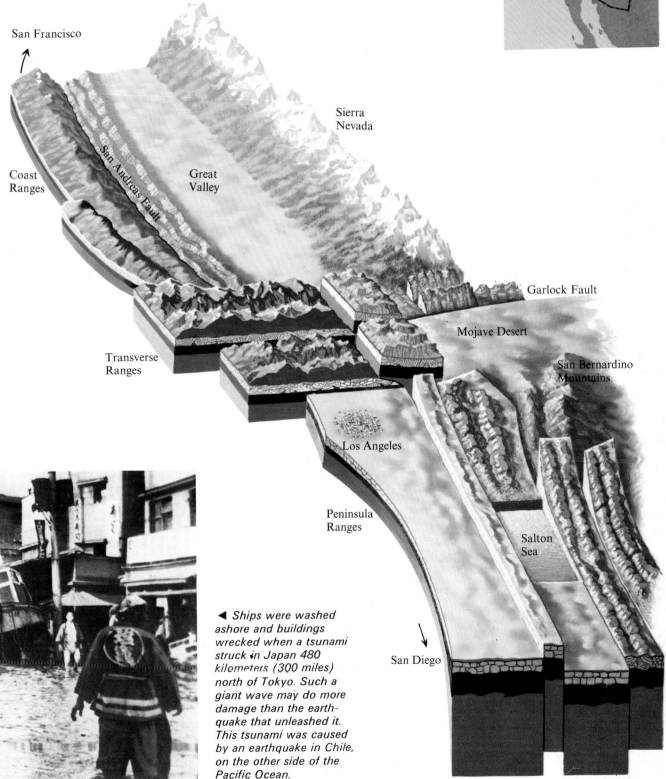

◄ *Ships were washed ashore and buildings wrecked when a tsunami struck in Japan 480 kilometers (300 miles) north of Tokyo. Such a giant wave may do more damage than the earthquake that unleashed it. This tsunami was caused by an earthquake in Chile, on the other side of the Pacific Ocean.*

# Mountains

If you were circling the Earth in a man-made moon, you would notice that much of the land looked like a wrinkled rug. The humps would be mountains, and the dips between them valleys.

Some areas are more mountainous than others. Maybe you could pick out two great rows of mountains. One runs east from northwest Africa and southern Europe through Asia. This line includes the European Alps and the Himalayas, separating India from China. Some of the highest mountains in the world are in the Himalayas.

A second massive row of mountains runs from north to south, close to the western coast of the Americas. This row includes North America's famous Rocky Mountains, and South America's great mountain chain, the Andes. The highest Andean peaks are almost as high as the highest Himalayan mountains.

As well as these mighty walls of rock, you might glimpse lower, shorter groups of mountains. For example, there are mountains on the edge of the Great Rift Valley of East Africa. The Great Dividing Range runs close to Australia's eastern coast. Antarctica is also mountainous. But a thick ice blanket hides all but its highest peaks.

## Fold Mountains

If you climbed to the top of Mount Everest, you might get a surprise. For a close look at its rocks would show the remains of creatures that once

▲ *A distant view of the Himalayas, the world's highest mountains. These great walls of rock were pushed up as India moved north and collided with the rest of Asia.*

▼ *This cutaway view through the Earth's surface shows some types of mountains and valleys formed when rock masses collide or slide up or down against one another.*

▶ *The edge of the East African Rift Valley in Kenya. The Rift Valley is 6437 kilometers (4000 miles) long, and runs along the Red Sea and through the countries of Ethiopia, Kenya, Tanzania, Malawi and Mozambique. The Rift Valley formed when part of the land sank, forming a vast trench, and leaving a flat plateau above. Much of the valley is below sea level.*

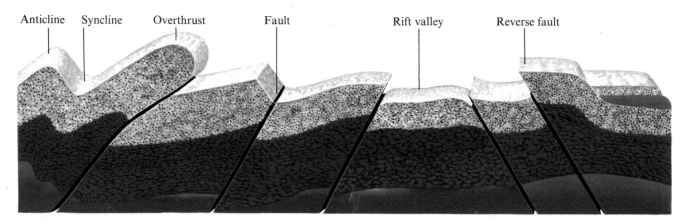

Anticline  Syncline  Overthrust  Fault  Rift valley  Reverse fault

▶ *Fossil animals in its rocks show that this mountain's limestone formed under the sea 250 million years ago. Movements of the Earth later forced it upward.*

lived under water. Indeed, the rocks belonging to the world's highest mountain were once at the bottom of an ancient sea.

In those days, India was an island separate from the rest of Asia. But about 50 million years ago, a drifting crustal plate bore India northward until it crashed into Asia. What had once been sea-bed rocks were squeezed between the two as someone shuts a concertina. Colliding rocks reared up. Some were folded over onto others. So, over millions of years, rose the mighty Himalayan platform.

In the same way, Italy and southern Europe crashed together and the Alps were born.

Movement of two crustal plates also helped to raise great mountain chains in the Americas. When North and South America moved west and hit the huge Pacific Plate, their western rim and parts of the nearby sea bed crumpled up to form the Rockies and the Andes. Meanwhile the eastern edge of the Pacific Plate dived beneath the two Americas.

## Block Mountains

The mountains we have just described are largely *fold mountains*—mountains thrown up when huge forces buckle rock layers into giant wrinkles. But movements of the Earth also produce *block mountains*. Block mountains may be formed where moving plates put strains upon the crust. Strains

| **Mountains—Facts and Figures** |
|---|
| The world's highest peak is Mount Everest. It is 8848 meters (29,028 ft) above sea level. |
| The world has 109 peaks above 7315 meters (24,000 ft). All of them are in Asia, and 96 of them are in the Himalaya-Karakoram Range. |
| The greatest of all mountain systems is the Mid-Atlantic Ridge, 16,100 kilometers (10,000 miles) long. Most of its peaks lie beneath the ocean. |
| The tallest active volcano is Cotopaxi in the Andes of Ecuador. It is 5897 meters (19,347 ft). |

like these also cause earthquakes. In places several parallel cracks or faults may open up in the land. A block of land between two such faults may be forced up by pressure from each side. The Vosges Mountains of France and Germany's Black Forest are mountain regions that were formed like this.

Elsewhere, a block of land may slip down between two others. It then becomes a *rift valley*. One such valley runs about 6437 kilometers (4000 miles) from Syria to southeast Africa.

## Lonely Peaks

Not all mountains build up in lines. Some of the largest are built up separately as volcanoes. Indeed the largest mountain in the world is a volcano. Mauna Loa in Hawaii rises 9750 meters (32,000 ft) from the ocean floor.

# Land Attacked

Land is built up by volcanoes and colliding continents. But as soon as land appears above the sea, it comes under attack. Sun, wind, frost, ice, chemicals and moving water combine to wear it down. Their work is slow.

It may take millions of years to wear a mountain down to the level of the sea. But everywhere at least one of these agents of destruction is busy. Between them, they have worked on all but the newest volcanoes, and shaped all other hills and mountains, and every plain and valley in the world today.

**Weathering at Work**

Attack upon the land begins with *weathering*. This is the rotting or splitting up of solid surface rock. Changing temperatures can split up rock in several ways.

In hot lands, the Sun heats up the surface rock, making it expand more than the rock below. The surface rock may flake off, much like the outer layer of an onion.

In cooler lands, rain may fill up the cracks in the rocks. On cold days or nights the water freezes. Because ice takes up more space than liquid water, it presses outward in the cracks and makes them larger. Freezing and thawing may

◀ Frost and ice have broken up the surface of the land. First, rain filled cracks and hollows in solid rock. On cold days and nights, this water froze. But ice takes up more space than water. The pressure of the ice split fragments from the solid rock. These slid downhill until the steep slope was surfaced by loose rocks and stones. Such slopes are called screes.

◀ This bare, bleak surface is a karst landscape. It takes its name from a part of Yugoslavia. It consists of limestone that has been attacked by rain. Rain falling through the air dissolves carbon dioxide gas. It makes a weak acid that dissolves limestone. The acid widens cracks in the rock and sinks through them, producing a barren, rocky landscape.

▶ These cliffs show where sea attacks the land. Wind-driven waves hurl water, air and stones against the foot of a cliff until the top collapses. Sometimes the sea cuts a cave through a headland. If the cave roof falls, the tip of the headland is left standing as an island in the sea. The islets in this photo once formed parts of the cliffs, most of which the sea has now eroded.

▲ *Wind-worn rocks in the Algerian part of the Sahara Desert. Wind hurls loose sand grains against the bases of the rocks. In time the rocks are undercut and topple.*

▶ *Sand makes wave-like dunes in many deserts. First, sand grains are torn from solid rock by sun, frost, wind or rain. Then the wind blows the sand away across the land.*

happen many times each year. Over the years, it breaks the rock surface into stones and boulders.

Breaking up surface rock by heating and freezing is called *mechanical weathering*. But rock is also broken up by chemicals. *Chemical weathering* takes place when water or chemicals contained in water actually dissolve or rot the rock away. For instance, rainwater rots the hard rock granite by changing feldspar (an ingredient of granite) to kaolin, a soft clay. Falling rain collects carbon dioxide gas from the air. The two form a weak acid that eats into the limy rocks, chalk and limestone. Some limestone hills have many caves gnawed out by water flowing underground.

## The Work of Wind
The shaping of the land begins when rocks are broken up, as in the kinds of weathering that we have just described. Before hills and valleys can be formed, loose rock material must first be removed. Nature's removal men are moving air, water and ice. Their shaping process is *erosion*.

In deserts, weathered particles of rock form sand. Windblown sands strike the bottom of rocks with stinging force. They undercut boulders, and gouge caves and hollows in the ground. Where winds blow steadily from one direction, the sands may pile up into shifting hills, called *dunes*. Winds not only erode loose rock, they carry it from one place to another.

## The Restless Sea
Sea is more powerful than wind as a weapon for destroying and creating land. But sea power is largely wind power at second hand. For the sea is most destructive when gales whip up its surface into waves.

A storm wave crashing on a sea cliff crams air into cracks in the rocks. When the wave falls back, the air expands. This process gradually widens the cracks and breaks off chunks of rock. The sea then hurls these rocks against the cliff. In time, the sea eats away the cliffs. Meanwhile, the sea breaks down broken rock into pebbles and sand grains, which are carried along to calmer water. There the sea drops its load. This builds up into sandy or shingly beaches. So new land is made from old.

# Rivers and Ice

> **Rivers—Facts and Figures**
> The world's rivers hold about 230,000 cubic kilometers (55,000 cubic miles) of water.
> The longest river is the Nile. It is about 6670 kilometers (4145 miles) long.
> The largest river is the Amazon. It pours 120,000 cubic meters (123,000 cubic yards) of water per second into the sea.

Running water and moving ice do most to change the surface of the land. Rivers have cut valleys from the uplands or dropped loads of mud and silt upon low-lying plains.

Many rivers are produced by rain. When rain falls on the land, much of it sinks into the soil. Some seeps down through rocks. Rocks which allow water to slip through them are called *permeable*. In places, a permeable rock layer lies above an impermeable layer. Water fills the permeable layer to a level called the *water table*.

## How Rivers Alter the Land

Many rivers start as springs high up on hills or mountains. From such a spring a small stream flows quickly down a steep slope. As it flows, the water picks up stones already broken from the surface rock by weathering. These moving stones rub on the stream bed and deepen it. They may form waterfalls and rapids too.

As the stream deepens its bed, it is joined by rivulets of rainwater flowing down its steeply sloping bank. These tributaries help the stream to grow into a river. Young, upland rivers carve out steep-sided valleys. A slice across one would look something like a letter V.

In time, a river will have carved a deep, broad valley from the land. The slope of the stream bed gets gentler. As the valley broadens, its sides become less steep, and the river winds along a valley floor in bends called *meanders*. The river's strongly flowing current hits and wears away the outer bends. But the flow is slack on inside bends. Here, the river drops a load of particles of gravel, sand and mud, largely made by water grinding stone on stone.

Near its mouth, a river may reach "old age". Here it has worn away hills and mountains. It winds slowly toward the sea over broad low, almost level land. This *flood plain* may end in an *estuary* or a *delta*.

Big rivers drop huge loads of mud and silt on the sea bed. Pressure from above squashes and hardens such *sediments*. In time they may become layers of *sedimentary rock*. Later still, movements of the Earth may raise these rocks above the waves. So the story of erosion starts again.

## The Work of Ice

Much of the world's fresh water is frozen. A huge ice cap covers Antarctica. *Glaciers* (rivers of ice) form from thick layers of snow and fill many mountain valleys. As they creep downhill, the glaciers pick up stones that deepen and widen their valleys. A slice across a glaciated valley looks like a letter U.

▶ Frost and ice sharpen peaks and enlarge valleys in high mountains. A lake forms behind rock debris dropped by a glacier. From the lake a fast-flowing stream cuts a steep-sided zig-zag valley. Lower down the river flows less fast and has carved a broader, gentler valley. Lastly, the river flows slowly in big loops over a low plain and drops mud into the sea.

◀ Where sluggish rivers meet the sea, they drop their loads of silt and mud. Mudbanks and islets begin to grow. New land is being made from old.

▶ Ice helps to shape many mountains. Rocks stuck in the ice of a glacier wear away the walls and floor of the valley it flows through. This deepens and broadens the valley. At the same time, the melting and freezing of snow and ice breaks up rocks at the top of the valley. This helps the glacier to eat back into the mountain peaks and steepen their slopes.

▲ The Colorado River cut into its bed, and carved the world's largest gorge. The Grand Canyon is nearly 350 kilometers (217 miles) long, and up to 2133 meters (7000 ft) deep.

# Chapter Three
# Life on Earth

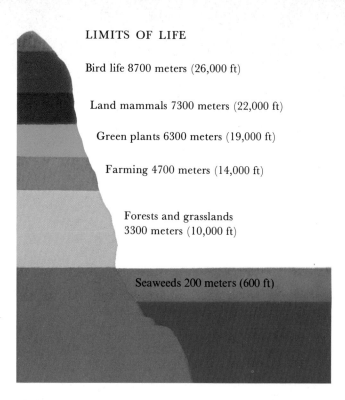

▲ *How high up or low down some types of plants and animals live. Green plants cannot survive the cold, thin air of the highest peaks. Lack of light stops them making food more than 180 meters (590 ft) under the sea.*

Very slowly, over millions of years, the Earth gained its cool outer crust, its oceans, and its air. Air, water, and the surface of the land make up the *biosphere*—the thin film of life clinging to the outside of our globe. There is no proof that living things exist anywhere else on Earth or anywhere beyond.

The film of life is very thin. Tiny windblown scraps of life, such as spores and pollen grains, may drift into the upper air. But birds fly no higher than the cold tops of the highest mountains. Mammals and green plants have to stay at lower levels. Forests and grasslands flourish lower still.

Farther down, only the upper levels of the sea have enough light for plants to grow. But animals live in the deepest ocean valleys, feeding on the dead and dying plants and animals that drift down from above. Beneath the surface of the land, a few kinds of plants and animals live in sunless caves. There is no life below the upper layers of Earth's crust.

Although the biosphere is thin, it teems with living things. The very air you breathe is often full of tiny organisms called bacteria and viruses. Some of these are too tiny to be seen by any ordinary microscope.

Living things range in size from these specks to the giant redwood trees—the largest living things on Earth. Some weigh more than 2000 tons.

Life also comes in unbelievable variety. Scientists have counted more than 300,000 kinds of plant and more than 1,000,000 kinds of animal.

Green plants are the food factories on which almost every other living thing depends. Only green plants can use the energy in sunlight to turn some of the chemicals in water, soil or air into food to help them live and multiply.

All animals obtain their food by eating plants or other animals. In the end, all living things depend on the plants' ability to make their own food. In any piece of land, plant-eating animals are usually the most abundant. Some kinds are astonishingly common. A chunk of meadow turf no longer than a man's foot may hide 5000 of the tiny insects known as springtails.

▶ *A family of* Rhesus *monkeys at home among Sri Lanka's forest trees. The warm wet tropics are rich in wildlife. But almost any place on the Earth's surface yields food and living space for some kind of plant or animal.*

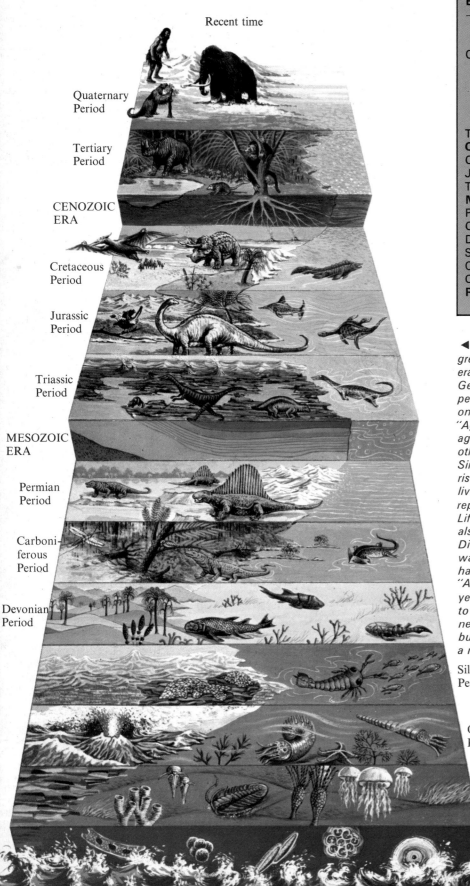

# THE DEVELOPMENT OF LIFE

| THE AGES OF THE EARTH | |
|---|---|
| Era, Period, Epoch | Millions of Years Ago |
| Pleistocene Epoch | 10,000 years –2 million |
| Quaternary Period | |
|   Pliocene Epoch | 2–5 |
|   Miocene Epoch | 5–25 |
|   Oligocene Epoch | 25–40 |
|   Eocene Epoch | 40–55 |
|   Paleocene Epoch | 55–65 |
| Tertiary Period | |
| **CENOZOIC ERA** | |
| Cretaceous Period | 65–135 |
| Jurassic Period | 135–200 |
| Triassic Period | 200–225 |
| **MESOZOIC ERA** | |
| Permian Period | 225–280 |
| Carboniferous Period | 280–345 |
| Devonian Period | 345–395 |
| Silurian Period | 395–440 |
| Ordovician Period | 440–500 |
| Cambrian Period | 500–600 |
| **PALEOZOIC ERA** | |

◀ How life on Earth has evolved in three great ages of Earth history, called eras. Each era makes up one step in the illustration. Geologists divide each era into periods. Some periods are split into epochs. Life flourished only in the sea when the Palaeozoic Era or "Age of Ancient Life" began 600 million years ago. By the time it ended, trees and many other plants were growing on the land. Simple animals without backbones had given rise to fish. From fish came amphibians, which lived on land and in water. From them came reptiles. The Mesozoic Era or "Age of Middle Life" saw reptiles rule the land and seas. It is also called the Age of Reptiles or the Age of Dinosaurs. The biggest beasts that ever walked on land flourished in this era, but they had died out by the time the Cenozoic Era or "Age of Recent Life" began about 65 million years ago. Meanwhile reptiles had given rise to birds and mammals. Mammals became the new rulers of the land. Man, too, is a mammal, but the first men like ourselves lived less than a million years ago.

▶ Fossils in layers of rock. Usually, the lower the layer, the older its fossils. But folds, faults and erosion can jumble up the layers, bringing old fossils to the surface.

▲ *The fossil of a giant fish. Back in Cretaceous times, it swam in a sea where Kansas stands today. Two men point to a six-foot fish swallowed by the giant before it died and sank into chalky mud.*

# How Life Has Changed

Locked in the rocks is the story of how modern plants and animals got where they are today. Millions of years ago, sedimentary rocks such as sandstone, shale and limestone were laid down on the floors of seas or lakes, as layers of sand or mud. Here and there the sand or mud trapped dying plants and animals. Their fossil remains are preserved in the rock and are clues to the kinds of life that flourished in those distant days.

In time one sedimentary layer was often covered by another, with a different set of fossils. By comparing the fossils in several sets, scientists have learned how living things *evolved*, or changed. Special tools also help them to date the rocks and the fossils.

The Earth took shape about 4600 million years ago. Scientists believe that life first appeared hundreds of millions years after that. In those days, the warm oceans were like a rich broth of chemicals. Sunshine playing on certain chemicals may have produced the first living things—tiny particles able to use food to grow and reproduce themselves. By 3300 million years ago, tiny plants were drifting in the oceans. But it was a long time before complicated kinds of plants and animals appeared.

Rocks show signs that modern life was beginning by about 600 million years ago. The seas began to teem with familiar living things, such as jellyfish and seaweeds. Later came sea animals with backbones, and plants and animals that lived on land. Birds and mammals evolved only much more recently.

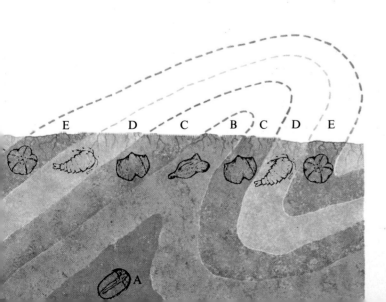

**KEY**
A Cambrian fossil
B Ordovician fossil
C Silurian fossil
D Devonian fossil
E Carboniferous fossil
F Permian fossil

▶ *Fossil leaves preserved in stone. They belonged to* Annularia. *This tree-sized relation of modern horsetail plants grew in swampy forests about 300 million years ago.*

# Plants and Climate

Most land plants need soil, warmth, light and moisture. But not all need the same amounts or kinds of each. Thus different kinds of plants grow in different climates.

Polar climates are too cold for plants except for low-growing, shallow-rooted plants such as lichens, mosses, grasses, shrubs, and tiny trees.

Mountain climates are cold and very windy. Only small plants can grow here. They crouch in rock crannies, or are shaped like cushions, to protect them from the bitter wind.

Conifers are the main trees that grow in cold forests. Many have sloping sides, and leaves like soft green needles. Snow can slip off such trees without damaging their branches.

▲ A treeless tundra scene. Only lichens and low-growing plants with shallow roots thrive in the cold polar regions of the world.

▲ Yellow flowers brighten a gray mountain slope in Kashmir, 3660 meters (12,000 ft) above sea level. Only low-growing plants can survive the fierce dry gales and intense cold of the world's high peaks and mountains.

## Plants of Temperate and Tropical Lands

Broad-leaved trees grow in lands with temperate climates. They are called deciduous trees because they shed their leaves in the fall. Evergreen trees and shrubs grow in the warmer parts of these regions. Their shiny, or leathery, leaves help to store moisture during the long, hot, dry months of summer. Cool temperate lands which are too dry for trees are covered instead by the rolling grasslands of the prairies, steppes and pampas.

Where there is very little rain, the soil is usually dry. Only plants like the spiny-leaved cactuses

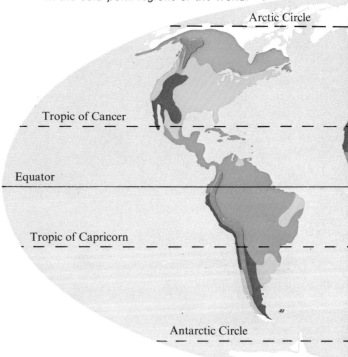

WORLD CLIMATES

- Polar (cold)
- Mountain (cold, windy)
- Cold forest (cold winters)
- Temperate (mild winters)
- Desert (dry)
- Tropical rainy (warm, moist)

and the fleshy-leaved succulents can live in such deserts.

Nearer the equator, in savanna country, there is just enough rain for scattered trees to grow among tall grasses. Parts of the tropics are very hot and very wet. Some of the world's greatest forests grow here. Their trees are tall, broad-leaved and evergreen.

**Look-Alike Plants**

Similar plants tend to thrive in similar climates, even if an ocean separates them. Some such plants are closely related. They evolved together before the sea opened up between them. But many plants just happen to look alike. They developed to suit the same climate.

▲ Part of a great forest of evergreen, cone-bearing trees. Coniferous forests ring the world's cool lands near to the tundra.

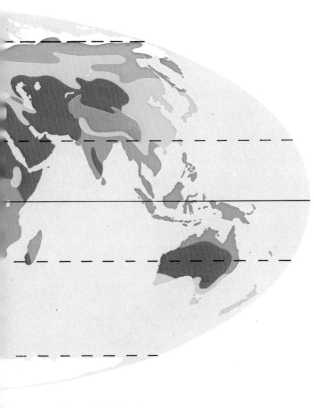

▲ Inside an oak forest. Its trees shed their leaves for winter. Such deciduous forests grew in many temperate lands near the coniferous forests. But most of the world's deciduous forests have been cut down by man.

◄ A bird's nest fern sprouts on the branch of a tree in a tropical rain forest in Madagascar.

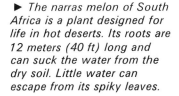

► The narras melon of South Africa is a plant designed for life in hot deserts. Its roots are 12 meters (40 ft) long and can suck the water from the dry soil. Little water can escape from its spiky leaves.

# Animals of the World

Like plants, many groups of animals are better suited to the climate of one region than to that of others. Some have stayed in one region because they evolved only after the sea or rising mountain ranges had cut off their region from the rest of the world.

▲ *Millions of bison used to graze on the temperate Nearctic (North American) prairies.*

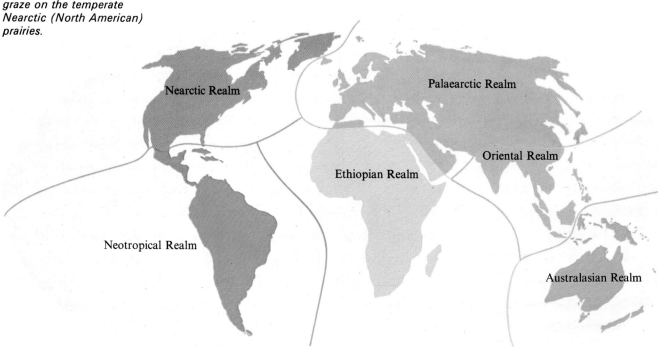

▼ *An anteater licks up insects with her tongue, while her baby takes a ride on her back. Anteaters, sloths, and monkeys that can hang by their tails belong to Central and South America (the Neotropical Realm).*

▼ *Zebras drinking at a waterhole in Africa. Zebras, antelopes, lions, gorillas, and ostriches are among the animals that live in parts of the Ethiopian Realm (Africa and Madagascar). Most animals there are suited either to hot forests, scrub, grassland or deserts.*

▲ *The common hedgehog is a mammal which belongs to Europe and northern Asia, part of the Palaearctic Realm.*

▼ *Gibbons live in the Oriental Realm, in south and south-east Asia.*

▼ *A platypus swimming under water. It is one of only two mammals that lay eggs. Both live in the Australasian Realm.*

Zoologists have divided the world into six main animal regions or realms.

Many zoologists now lump the Nearctic and Palaearctic realms into one, Holarctic, realm. Holarctic animals include the perch, salmon and sturgeon—fish found nowhere else. More beasts are special to the Nearctic than to the Palaearctic area. Nearctic animals include the North American antelope (the pronghorn), two lizard families, and the bowfin, an ancient type of freshwater fish.

The Neotropical Realm of Central and South America has an even richer store of animals peculiar to one realm. New World monkeys, sloths, anteaters, and many rodent families are among its special mammals. Nowhere has so many kinds of birds as South America. About a dozen families of freshwater live only in the Neotropical Realm.

There are many animals which are special to the Ethiopian Realm (Africa and Madagascar). Its unique mammals include giraffes, hippopotamuses, the aardvark, and a group of lemurs.

## Oriental and Australasian Realms

The Oriental Realm—the smallest—has many strange beasts, such as tree shrews, flying lemurs, and the spiny dormouse.

But the strangest animals of all inhabit Australasia. This region is the home of most marsupial mammals—such beasts as kangaroos and koalas, whose mothers raise their babies in a pouch. Here, too, live the platypus and the echidna—the only mammals that lay eggs.

## Gains and Losses

In each animal realm, man has made some species extinct, but brought in others from outside. The Nearctic Realm has lost the passenger pigeon, heath hen and eastern bison. The Palaearctic Realm now lacks its aurochs (a wild ox) and tarpan (a wild horse). The Ethiopian Realm no longer has its zebra-like quagga, its dodo, or *Aepyornis* (a giant bird).

The Oriental Realm has lost its Balinese tiger. For the Australasian Realm, New Zealand's great bird, the moa, is just a memory. The giant ground sloth and glyptodont (a giant armadillo) of the Neotropical Realm are also gone for ever.

However, men have shipped domestic cattle, sheep and fowls to almost every country of the world. Besides these farmyard animals, crop pests, including rabbits and house sparrows, have crossed the oceans.

# Chapter Four
# Man on Earth

▶ *Oxen help a peasant to plough a rice field in India. All over the world, farmers produce most of the food we eat. The Earth now feeds more than 4000 million people, and will soon have to feed many more.*

The direct ancestor of man was a kind of ape, now long extinct. People proved to be different from other animals. They developed languages which helped them to think and reason. Their big brains and nimble fingers helped them make tools. They used the world to feed, clothe and house themselves. They found out how to make fire to warm themselves and cook their food. They also invented the wheel to help them move loads about.

People spread around the world, probably from Africa. In time, people in different lands evolved in different ways. Negroes have dark skins to protect them from the hot sun's harmful ultraviolet rays. Caucasian (white) people have paler skins to help them absorb as much weak sunlight as their bodies need. Mongoloid peoples have shorter arms and legs to help them to keep in body heat in cold climates.

Whatever their race, almost all people owe more to one invention than to any other. That invention (or discovery) was farming. Once they learned how to grow crops and to raise animals, they gained a surer supply of food. Also, one farmer could feed several people. Today, farmers produce extra food for miners, factory workers, traders, and many other people. All work together to make the world's wealth serve the needs of mankind.

▼ *The world's main languages. People in different lands developed into races, with different languages, religions and societies.*

MAIN LANGUAGES
- Indo-European
- Sino-Tibetan
- Mon-Khmer
- Japanese and Korean
- Uralic and Altaic
- Dravidian
- Malayo-Polynesian
- African
- Afro-Asiatic
- Other

# Using the Rocks

From left to right: a metamorphic rock, two igneous rocks, two sedimentary rocks. Each main type of rock provides building stone or other substances.

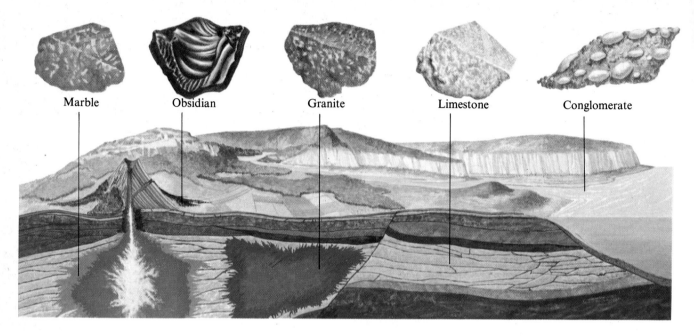

Marble — Obsidian — Granite — Limestone — Conglomerate

Farming and fishing provide people with food. Farms supply useful goods like wool and cotton, too. Forestry provides timber for buildings, furniture and paper. But many substances are quarried or mined from the rocks or soil. Building stone, for example, usually has to be quarried.

## Rocks as Building Blocks

Some types of building stone consist of *igneous*, or "fiery" rocks. One of these is granite. Granite is a hard rock that may be formed when molten rock cools underground.

Other kinds of building stone come from *sedimentary* rocks, such as sandstone and limestone. Both were laid down as sediments beneath prehistoric seas and lakes. Limestone rocks are ground and burnt to make cement. Builders make concrete by mixing cement with sand and gravel —two more kinds of sedimentary deposit. Certain kinds of melted sand are used for making glass.

Bricks are blocks of clay that have been baked in ovens which are known as kilns. Bricks are usually used where there is no ready supply of local building stone.

Yet other kinds of building stone come from *metamorphic* rocks. This third great group of rocks consists of rocks changed by nature, under immense heat or pressure. Slate is a metamorphic rock that was once shale. Marble is a very hard metamorphic rock, which is sometimes used for floors and walls.

## Minerals, Ores and Metals

Besides stone for building, rock provides many useful substances such as the talc for talcum powder, graphite for pencil "leads", and fertilizer for the field and garden. Of the 2000 *minerals* contained in rocks, the most useful are perhaps the *ores*—the minerals containing metals. Metals, such as iron, copper, aluminum, are vital ingredients in many modern tools and machines. Unfortunately some useful ores are scarce or difficult to find.

## Putting Minerals to Work

Prospectors now use advanced technology to help them discover new mineral deposits. Seismographs and gravimeters give clues to rocks deep underground. Magnetometers locate metals. Geiger counters reveal radio-active substances.

Once a deposit has been found it may be mined or quarried. If it lies just below the surface, men may strip away the soil and dig it out with huge power shovels. But quarrying or open-cast mining is not always possible. Then miners may drill or tunnel deep down in the rock. When miners have obtained an ore, the metal in it must be separated from unwanted substances. For instance, getting iron from its ore involves smelting—heating the ore to a high temperature in a blast furnace.

From metals purified like this or mixed as alloys, factory machines can mold or stamp out engines, car bodies, and other products.

▲ An Australian iron-ore mine. This hill is so rich in iron ore that men do not have to tunnel for it. Instead their machines remove the whole hill, bit by bit. Giant steps show where huge slices have been cut out and carted off.

▼ Testing molten iron inside a blast furnace. Great heat is needed to separate iron from its ore.

▼ Making cars in Japan. Iron is made into steel to provide millions of car bodies.

# Energy

Rocks provide more than building blocks and materials for manufacturing machines and tools. Most of the energy used for light, heat, and power is obtained by burning fuels dug up or piped from underground.

## Fossil Fuels and Wood

Three fuels are especially important. These are coal, mineral oil and natural gas. Most coal was formed in Carboniferous times. Forest trees were drowned by sea and squashed by sediments that piled up on their remains. Oil and gas were also formed many million years ago, by tiny organisms trapped deep underground. Coal, oil and gas are called *fossil fuels*.

Some shallow coal deposits can be scooped up from the surface. But British miners in one colliery had to dig down 1259 meters (4132 ft) to reach coal. Prospectors have drilled far deeper down

◀ Waterfalls like this unleash enormous quantities of energy. That energy can be used to spin wheels and generate electricity. Such hydro-electric power plants have been built at many major waterfalls.

▼ Drilling for coal deep underground in dark, cramped and dangerous conditions. Coal has been used to heat homes and drive machines for hundreds of years.

than that to find oil and natural gas trapped in the rocks. There is usually enough pressure underground to force the gas or oil through a pipe up to the surface, but sometimes oil has to be pumped to the top. The different substances in oil are separated in oil refineries to give kerosene, gasoline and other fuels.

Coal, oil and gas can all be burnt to produce electric current. The burning fuels turn water into steam. The steam spins wheels whose movement generates electricity.

Oil and coal are also used to produce most kinds of plastic sheets and fibers.

Besides these fossil fuels, some people burn other fuels that were produced by living organisms. *Peat* is a fibrous mass of moss and other plants that died in bogs. In some places *wood* is still an important fuel.

## Water Spins a Wheel

All the fuels so far described store energy obtained by plants from sunlight. The water in a lake or river also holds energy, thanks to the Sun. The warmth from the Sun causes the water to evaporate from the sea and land, later to fall as rain and run downhill in rivers. Falling water can be made to spin a wheel which can be used to work machinery. People were using waterwheels to grind grain many hundred years ago. Today, falling water is mainly used to spin turbines and generate electric current. Electricity produced like this is called *hydroelectricity*.

## Energy from Atoms

Uranium has far more energy locked up in it than any source of power so far described. One kilogram yields nearly as much energy as 4,500,000 kilograms of coal. To tap the energy in uranium atoms, the uranium is mined and used to power nuclear reactors. There, the immense heat given off by certain types of uranium is used to boil water and produce steam to generate electric current for homes and industry.

▶ Oil and natural gas is obtained from holes drilled deep into the ground. Escaping gas is burnt off at the surface where the oil and gas are drilled.

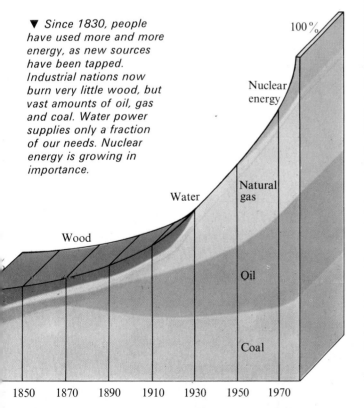

▼ Since 1830, people have used more and more energy, as new sources have been tapped. Industrial nations now burn very little wood, but vast amounts of oil, gas and coal. Water power supplies only a fraction of our needs. Nuclear energy is growing in importance.

# How People Live

► A busy scene in the city of Chicago. Mechanized farming means that 1000 farmers can feed 100,000 city dwellers. Workers in city factories and shops in turn provide goods and services. Today, hundreds of millions of people live in the world's great cities. Many people feel that city life creates too much strain

Farms, mines and factories have changed the ways that people live. Once, most people were farmers, producing food and living on the land. Now one farmer, helped by modern know-how and machines, can feed large numbers of people. There is no need for everyone to work on the land. Many people leave the land and make their homes in cities. They work in factories, offices, shops, schools, hospitals, trains or buses. One worker does only one kind of job. But everyone may benefit from it.

Many people now have cars, television sets and many other goods their ancestors never dreamt of —as well as nourishing, abundant food, and modern hospitals and medicines to cure disease. Most people in industrialized countries now lead longer, more comfortable lives than people long ago enjoyed. But even today not everyone lives in such comfort.

Some countries are poor. They cannot produce enough food to feed their people, and they have very little industry to provide wealth. Most people living there are also poor. Most die young from hunger or disease. In parts of Africa, most people die before they are 26. In the United States people usually live more than twice as long as that.

## More and More People

The population of the world is rising fast. In fact the number of people has been increasing for the last 10,000 years or so. The world fed perhaps a mere five million people in the Old Stone Age, when people relied on hunting for food. In the New Stone Age farming helped to feed four times that many mouths.

Trade between the farmers and the first cities led to another jump in population. There were perhaps 300 million people 1000 years ago. In the 1800s the total passed 1000 million. Now, more than 4000 million people live on Earth. If numbers go on rising at this rate the population of the world may more than double in your lifetime. The world's fast-growing population is causing many problems, as the next two pages try to show.

▶ *How the population of the world has increased, as people have found better ways of growing food and curing sickness. If numbers go on increasing at today's rate, the population will double in less than 40 years.*

▼ *Small boats bring fruit to a busy market in Thailand. Many Asians grow food for a living, but millions there are underfed and starving. Already more than half the people in the world live in Asia, and the population is increasing faster there than anywhere else.*

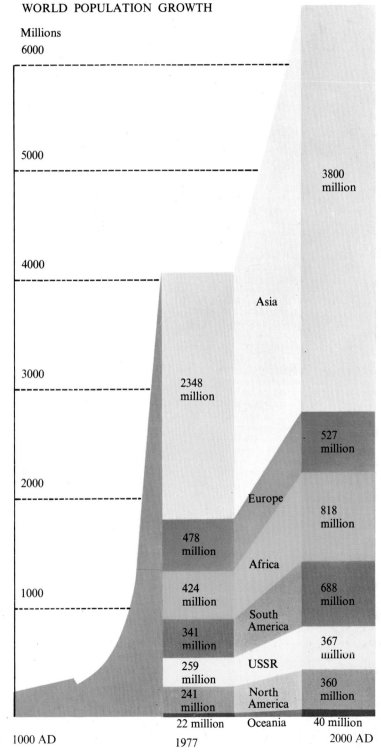

# The World Laid Waste?

As the number of people increases, so does their need for food, water, minerals and living space. To satisfy these needs, people have felled forests, drained swamps, and killed hundreds of kinds of animals, as pests or as food.

For centuries, the Earth has had more than enough plants, animals, rocks, and water to feed, clothe, house and warm mankind. But now the number of people strains these natural resources. In some places, farmers have wasted the soil and allowed it to be blown away.

Even where the soil stays fertile, much of it is used for raising sheep or cattle. A steer needs to eat many kilograms of grass to make one kilogram of beef. The land could be used instead to grow crops that people eat.

The number of some kinds of fish in the sea is getting less, because so many fish are being caught for food. Fish, fresh water and soil may all be renewed in time. But resources such as minerals and fossil fuels are also being used up, and they cannot be renewed. Before long, substances such as oil, coal, lead, zinc, tin and gold could run out altogether.

◀ *Factory waste pollutes a river, perhaps poisoning or suffocating its plants and animals. Wastes from cities, factories and farms threaten water, land and air.*

▼ *Cars, and many other factory products are soon worn out and scrapped. But some ingredients can be processed and re-used. Metal from this dump may make new cars.*

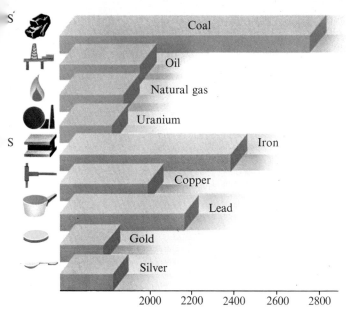

Amount of time stocks of some fuels and metals may last

▲ These materials may be used up by the dates shown if they are mined at their present rates. Unlike animals and plants, minerals cannot be renewed. As they get scarce, energy supplies and factory goods may run out.

▲ A solar energy plant in the Pyrenees. The Sun's rays are collected on the mirrors, and reflected on to a heat machine. This new way of tapping the Sun's energy may become more and more important in the future.

▼ Nuclear bombs like this could kill us all. But nuclear power stations may help supply tomorrow's energy.

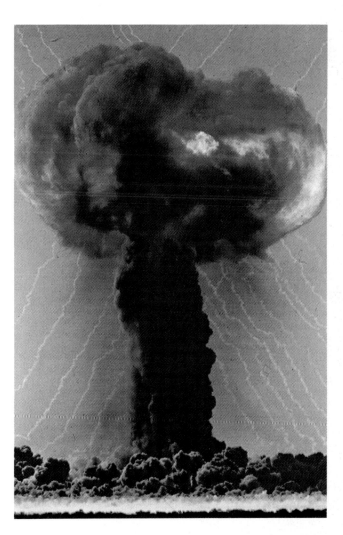

People are using up Earth's natural resources fast and wastefully. At the same time they are producing chemicals that damage the land, water, air or living things. On land, they have buried huge areas of farmland with concrete and tarmac for their cities and roads. There are great rubbish heaps on the edge of cities. Sewage, oil, factory wastes and pesticides find their way to rivers, lakes and seas. They poison the water and kill the plants and fish that live there. Fumes from car exhausts and smoke from factories pollute the air above some cities.

## Tomorrow's World

Earth's natural wealth helped people to become more successful than animals. But if they continue to mistreat the Earth, they put themselves at risk. Soon, even the rich countries could face hunger, poverty, disease. To prevent such a disaster, people must change.

They must have fewer children to stop the population growing so quickly. They should re-use waste metals, eat more plant foods, and less meat. They may even have to ration fuel supplies. In other words, we should *conserve* the natural resources we have. We must find new ways of getting energy—perhaps from wind or waves. Life tomorrow may be difficult. It could be exciting. It will certainly be different from today.

# Chapter Five
# Exploring Space

Out in space lie explanations for many things that happen here on Earth. The tides, day and night, the months, the seasons, the year—all of these are caused by other objects in the sky.

People have studied the Moon, Sun, and stars for thousands of years. Priest-astronomers used the Moon to make the first calendars. Sailors learnt how to use the Sun and stars to guide them when they were out of sight of land. Astrologers believed that they could tell the future from the positions of the stars and planets.

Yet many centuries passed before men learnt the truths about those pinpricks of light that twinkle in the night sky. Man's understanding of the heavens grew fast once telescopes had been invented nearly four centuries ago. Such instruments taught him much about the shapes, sizes, substances and journeys of the stars and planets.

But it was less than thirty years ago before anyone launched an object out of the Earth's atmosphere. Now people as well as unmanned spacecraft can speed out into space and send back fascinating information.

▶ *Far right. The Horsehead nebula, one of the most striking sights in the night sky.*

## LOOKING OUTWARDS

▼ *A copy of the first reflecting telescope, made by Isaac Newton in 1668.*

▼ *The world's largest refracting telescope at Yerkes Observatory near Chicago.*

### Some Famous Astronomers
Aristarchus (3rd–2nd century BC) suggested that the Earth spins and orbits the Sun.
Hipparchus (about 130 BC) listed above 1000 stars.
Nicolaus Copernicus (1473–1543 AD) revived Aristarchus' forgotten theory that the Earth orbits the Sun.
Tycho Brahe (1546–1601) made observations which showed that planets orbit the sun in elliptical paths.
Galileo Galilei (1564–1642) used a telescope to find out new facts about planets.
Isaac Newton (1642–1727) discovered the law of gravitation.
William Herschel (1738–1822) charted thousands of star clusters and galaxies.
Friedrich Bessel (1784–1846) was the first person to measure the distance to a star.
Ejnar Hertzsprung identified giant and dwarf stars in 1905.
Edwin Hubble (1889–1953) showed that the universe is expanding in all directions.

# Rockets

▼ One of the most advanced rockets of all. Three rockets thrust a manned satellite out of Earth's gravity and into orbit. Each rocket fires in turn. Only this "piggy-back" ride can thrust the satellite high enough to stop it falling down again.

▼ Saturn V's huge first-stage rocket engines yield a combined thrust of no less than 3½ million kilograms (7½ million lb).

▶ Cutaway view of the three-stage rocket system of Saturn V, used for the American manned Apollo flights.

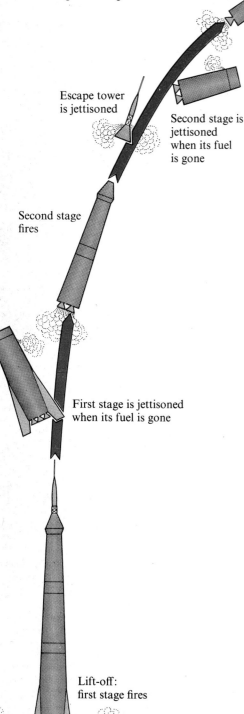

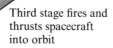

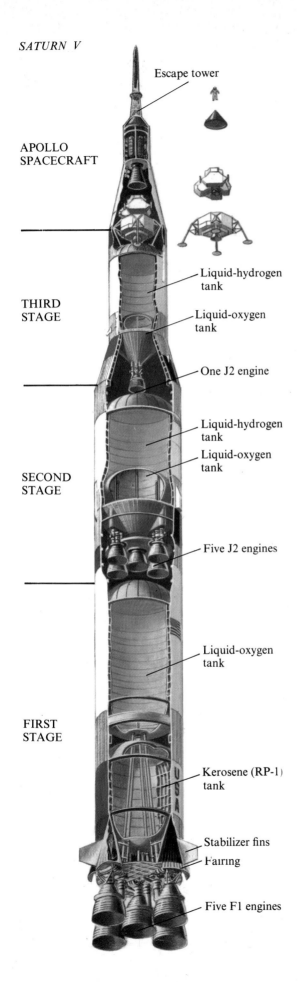

Putting any man-made object into space once seemed impossible. There were three reasons for this. First, the Earth acts like a giant magnet, holding objects to its surface. Escaping Earth's gravity called for an engine able to fly up into space at 40,000 kilometers (25,000 miles) an hour. The second problem was the lack of oxygen. On Earth, most ships, planes and cars burn fuel in oxygen drawn from the air. But space lacks oxygen. Thirdly, vehicles on Earth move forward by thrusting back against land, air or water. But space is empty of all three.

Yet children had long played with a type of engine capable of solving all three problems. This engine was the rocket. Firework rockets contain their own fuel and oxygen supplies. Gases given off in burning rush backward. This drives the rockets forward. Rocket engines can work in empty space. But the gunpowder in fireworks is hard to control and is not powerful enough to escape Earth's gravity.

In 1926 an American, Robert Goddard, helped to solve these problems. He built a rocket powered by a liquid propellant. This gave extra thrust and made his rocket easy to manage. In World War II German V2 rockets soared almost 100 kilometers (60 miles) high. By the 1950s, America and the Soviet Union had "piggy-back" rockets fast enough to zoom off into space or to set satellites (man-made moons) circling the Earth.

**Steps into Space**
1957 October, USSR launches *Sputnik 1*: first artificial satellite to orbit Earth.
1957 November, *Sputnik 2* puts a dog into orbit.
1958 January, *Explorer 1*, first US satellite, discovers Van Allen radiation belts.
1958 March, US *Vanguard 1* proves that the Earth is slightly pear-shaped.
1958 December, US *Project Score* broadcasts the first spoken message beamed from space.
1959 January, USSR's *Luna 1* passes the Moon and becomes the first man-made planet to orbit the Sun.
1959 February, *Vanguard 2*: first man-made moon to beam down weather information.
1959 September, *Luna 2*: first probe to hit the Moon.
1959 October, *Luna 3* photographs the hidden far side of the Moon.
1960 April, *Tiros 1* takes the first detailed pictures of the world's weather.
1960 April, *Transit 1B* becomes first navigational satellite.
1960 August, *Echo 1*: first communications satellite.
1960 March, *Pioneer 5* reports from 35 million kilometers (22 million miles) out in space.
1961 April, Yuri Gagarin completes the first successful manned space flight, orbiting Earth once in the Soviet spacecraft *Vostok 1*.

# Satellites

▶ *This world map shows something of the careful ground preparations needed to put artificial satellites in orbit around the Earth, and then to plot their paths. Rockets lift the satellites aloft from launching sites. Special equipment at the tracking sites finds the speed and distance of the spacecraft passing overhead. Besides the fixed sites shown, the United States and the Soviet Union can track satellites by means of special ships that range the oceans.*

▼ *Sputnik 1 was the first man-made moon to orbit the Earth. It weighed 83.6 kilograms (184.3 lb) and measured 58 centimeters (22.8 in) across. It was launched from Tyuratam in the USSR on 4 October 1957.*

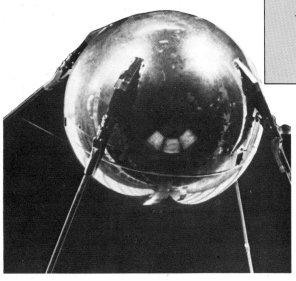

▼ *The eight giant "saucers" of a Soviet deep-space tracking station in the Crimea. These sensitive antennae can pinpoint the location of distant spacecraft.*

In October 1957 the Soviet Union amazed the world by sending up the first man-made moon. *Sputnik 1* (Fellow Traveler 1) was no bigger than a large party balloon. But it looped around the world at more than 28,000 kilometers (17,500 miles) an hour. At times it swung about 960 kilometers (600 miles) out. At other times it swooped down to a mere 227 kilometers (142 miles). As it looped, it radioed coded news of temperature and pressure changes.

Since 1957, the Soviet Union, the United States and some other countries have sent up thousands of artificial satellites. The instruments inside these man-made moons and space probes are artificial "eyes" that have told us much about the Earth and objects far beyond.

## Spies in the Sky

Less than four months after *Sputnik 1* took off, the United States launched *Explorer 1*. This man-made moon discovered the Van Allen Belts—two belts of radiation around the Earth. In 1958, *Vanguard 1* told its launchers that the Earth is not quite round but shaped more like a pear.

The same year *Sputnik 3* started work as the world's first complicated unmanned space laboratory. This satellite studied the make-up and the pressure of Earth's outer atmosphere. It also

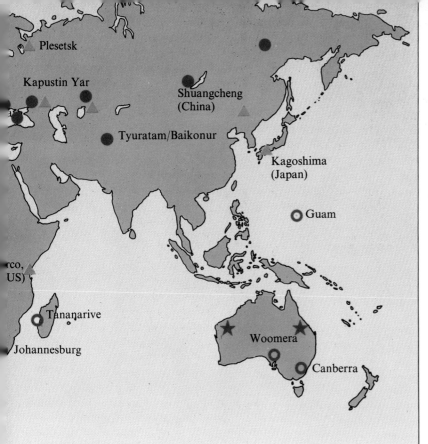

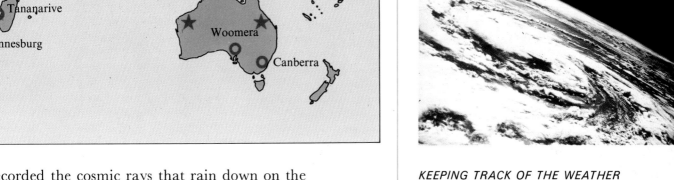

KEEPING TRACK OF THE WEATHER
Spies like this Nimbus *weather satellite can photograph the Earth's cloud cover.*

recorded the cosmic rays that rain down on the Earth from space.

In 1959 America's *Vanguard 2* became the first weather satellite. It beamed down pictures of the clouds above the Earth. Such satellites help meteorologists foretell the weather. Today's robot spies in the sky can map the entire Earth, and show places worth exploring for their mineral deposits. Meanwhile communications satellites relay radio, telephone and television messages around the world. Navigation satellites assist ships and planes to pinpoint their positions accurately.

**Unmanned Explorers**
Perhaps the most exciting news comes from robot spacecraft that leave Earth behind and probe into space. In 1959 the Soviet Union and the USA sent out the first space probes. *Lunar 1* and *Pioneer 4* beamed information from 35 million kilometers (22 million miles) out. Further robots have told us many facts about the Sun and planets (see later chapters in this book).

Only complex, delicate tools make all this possible. For instance, scanners keep satellite cameras pointed in the right directions. Solar cells trap sunshine to power radios.

Once engineers had sent up robot moons, they tried to put a man up there too.

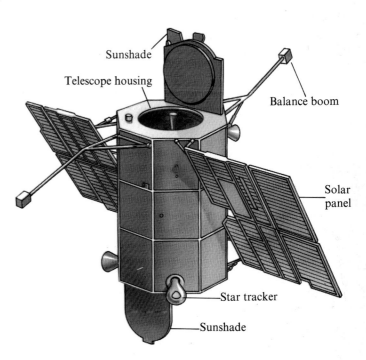

▲ *This satellite is an orbiting astronomical observatory (OAO). Space technologists have put several into orbit round the Earth. Up above the atmosphere their cameras observe the stars unhindered by air and clouds.*

# Man in Space

▶ *Yuri Gagarin and the* Vostok *in which he orbited the Earth in 1961. His pioneer spaceflight ringed the Earth in 108 minutes.*

During the 1960s the Soviet Union and the USA vied to set the first man in Earth orbit. Both countries knew that people in space could tell us more about the universe than instruments alone. But one problem was building a rocket powerful enough to lift a human passenger into space. Another problem was keeping him alive and well.

## Braving the Dangers of Space

Man's body is designed to breathe Earth's atmosphere, to move under atmospheric pressure, to resist the downward tug of gravity, and to live in comfortable warm surroundings.

Space has none of these. No life can live there without protection. Just 10 kilometres up, the air is too thin to breathe. At 20 kilometers deadly bubbles form in the blood. Twelve kilometers higher still the temperature is −55° Centigrade. At 38 kilometers the effect of ultraviolet rays is dangerous. At an acceleration of 11 kilometers per second (the speed required to break away from Earth) a spaceman's body weighs many times more than normal. When his rocket stops accelerating, he weighs nothing. And unprotected by Earth's atmosphere he could be hit by meteoroids.

Engineers found ways of saving astronauts from all these dangers. To house the spacemen they invented capsules strong enough to survive hits by tiny meteoroids. The capsules could also resist the fierce heat caused by air friction as the capsules hurtled back to Earth. Air inside each

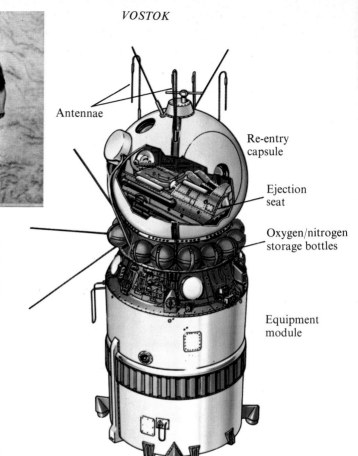

*VOSTOK*

▼ *The* Mercury *capsule took single astronauts into Earth orbit in America's first manned spaceflights.*

▶ *Edward White became the first American to walk in space in 1965, after an earlier Soviet space-walk.*

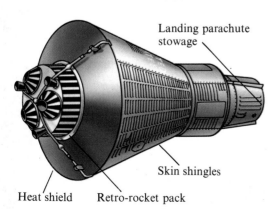

MERCURY

outer capsule protected an inner capsule from heat and engine noise. The inner capsule held warm, breathable air and a couch. Lying on it helped the astronaut survive the strain of fast acceleration. A space suit was his innermost defence. This supplied the air, pressure, and temperature he needed, as well as two-way radio equipment to help him keep in touch with Earth.

## The First Astronauts

Space suits, capsules and powerful rocket engines at last made manned space flight seem possible. But until 1961 no one knew if it would really ever work. That year Major Yuri Gagarin, of the Soviet Air Force hurtled into space in *Vostok 1*. Gagarin circled the Earth and landed two hours later. It was the fastest trip around the world that anyone had ever made. In 1962 John H. Glenn became the first American to orbit Earth by man-made satellite.

At first, the Russians led the way in manned space flight. In 1964 their spacecraft *Voskhod 1* put three men in orbit. But in 1967 Cosmonaut Vladimir Komarov died coming back to Earth: his spacecraft's parachute failed to open.

Meanwhile America was working hard to put men on the Moon by the year 1970. By 1963, six astronauts had circled the Earth in one-man *Mercury* spacecraft. In 1965 began the twelve two-man flights of *Project Gemini*. In 1968 the immense engines of the rocket *Saturn V* began thrusting three-man teams toward the Moon. *Project Apollo* (see page 59) was under way at last.

The United States led the way in manned Moon flights. But Russia and the USA have both sent up manned Earth-orbiting laboratories.

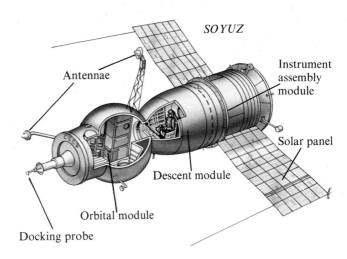

▲ *Cutaway view of a Soviet* Soyuz *("Union") spacecraft. In 1967, Vladimir Komarov orbited Earth in* Soyuz 1, *but his parachute failed to open as he returned. Komarov was the first astronaut known to have died in flight.*

### Steps to the Moon
1961 May, Alan Shepard becomes the first American spaceman, in a brief, suborbital flight.
1962 February, John H. Glenn becomes the first American to orbit the Earth (in *Friendship 7*).
1965 March, Virgil Grissom and John Young orbit the Earth in a two-man *Gemini* spacecraft.
1968 October, *Apollo 7* puts three astronauts in orbit around the Earth.
1968 December, Frank Borman, James Lovell, and William Anders orbit the Moon in *Apollo 8*.
1969 March and May, Lunar module successfully tested by *Apollo 9* and *Apollo 10*.
1969 July, *Apollo 11* puts Neil Armstrong and Edwin Aldrin on the Moon.

▼ *Frogmen secure the manned spacecraft of a* Skylab *mission after splashdown in the sea. During 1973 and 1974, three teams of American astronauts visited an orbiting space laboratory to study medicine, astronomy, and Earth's resources.*

▼ *This cutaway view of the three-man* Apollo *spacecraft shows the crew or command module, and the service module.*

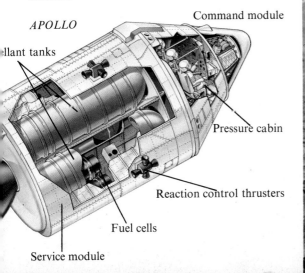

# Chapter Six

# The Moon

The Moon is Earth's nearest neighbor, and its only natural satellite. It is a huge ball of rock, somewhat like the Earth but smaller. The force of gravity exerted by the Earth prevents the Moon from speeding out into space. But it is always moving. It revolves around the Earth once every month. At night, the sunlight reflected from its surface shines palely down upon us.

The Moon has always fascinated man. Men worshiped the Moon as a kind of goddess who shed light at night. But some believed that too much moonlight drive men mad. From *luna*, the latin word for Moon, comes "lunacy", another word for madness.

For centuries people wondered what kind of world the Moon might be. In 1865, the French novelist Jules Verne wrote a story about man setting foot upon the Moon. The fantasy became fact 104 years later. In 1969 a rocket landed two Americans upon the Moon's surface. The rocks and photographs that they and later Moon explorers brought back have given a fuller picture of Earth's nearest neighbor.

▶ *The Moon as spacemen may see it. Even from Earth, its bare, bleak peaks and craters can be clearly seen. Yet only in the last few years have space probes and manned Moon flights revealed what kind of world Earth's nearest neighbor is.*

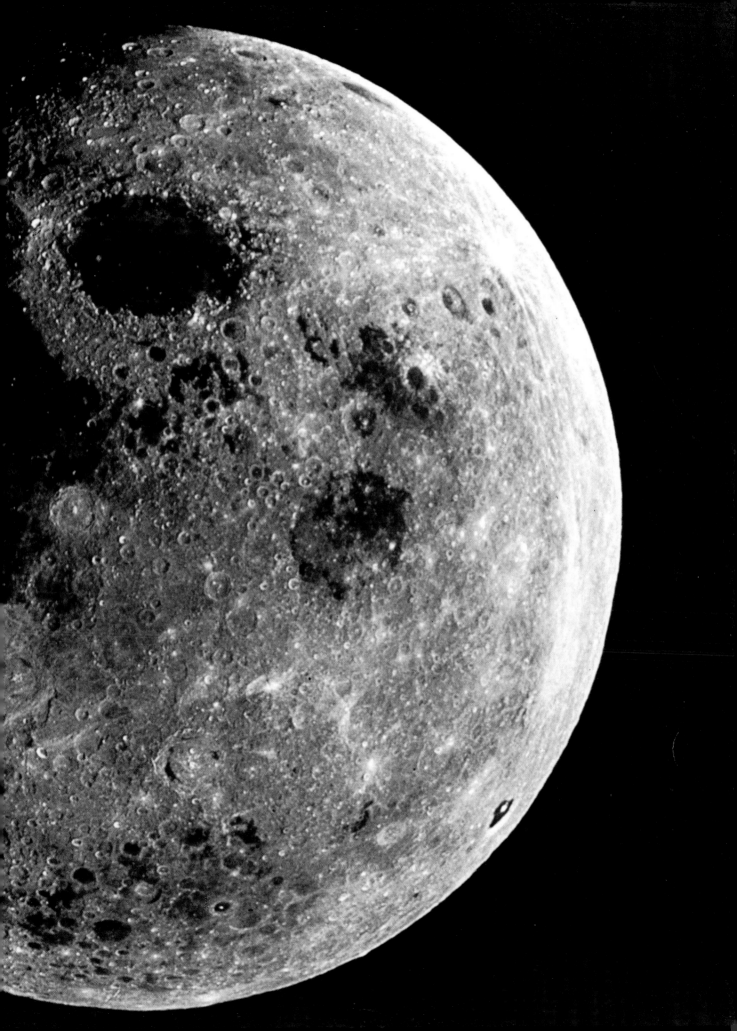

## Apollo Moon Landings

**Apollo 11** (16–24 July 1969)
Crew: Neil Armstrong, Edwin Aldrin, Michael Collins.
Command Module: *Columbia*; Lunar Module: *Eagle*.
*Eagle* landed on the Sea of Tranquillity on 20 July. Neil Armstrong and Edwin Aldrin spent more than 2½ hours on the surface. They collected 21 kg (46 lb) of stones and dust, and put a message and scientific instruments on the Moon. The world's largest-ever television audience watched them.

**Apollo 12** (14–24 November 1969)
Crew: Charles Conrad, Alan L. Bean, Richard F. Gordon.
Command Module: *Yankee Clipper*; Lunar Module: *Intrepid*.
*Intrepid* landed in the Ocean of Storms, only 200 meters (650 ft) from *Surveyor 3*, an unmanned probe sent there 2½ years before. In 8 hours, Bean and Conrad set up the first automatic scientific station on the Moon and gathered 34 kg (75 lb) of rock.

**Apollo 13** (11–17 April 1970)
Crew: James A. Lovell, John L. Swigert, Fred W. Haise.
Command Module: *Odyssey*; Lunar Module: *Aquarius*.
An explosion 320,000 km (200,000 miles) from Earth forced the crew to abandon the mission and return.

**Apollo 14** (31 January–9 February 1971)
Crew: Alan B. Shepard, Edgar D. Mitchell, Stuart A. Roosa.
Command Module: *Kitty Hawk*; Lunar Module: *Antares*.
*Antares* landed near the Fra Mauro crater. In two Moon walks, totalling 9½ hours, Mitchell and Shepard gathered 44 kg (96 lb) of Moon rock.

**Apollo 15** (26 July–7 August 1971)
Crew: David R. Scott, James B. Irwin, Alfred W. Worden.
Command Module: *Endeavour*; Lunar Module: *Falcon*.
*Falcon* landed near Hadley Rille in the foothills of the lunar Apenines. Irwin and Scott spent 18 hours on the Moon, gathered 76 kg (168 lb) of rock and explored 28 km (17 miles) in an electric "car".

**Apollo 16** (16–27 April 1972)
Crew: John W. Young, Charles M. Duke, Thomas H. Mattingly.
Command Module: *Casper*; Lunar Module: *Orion*.
*Orion* landed on the Cayley Plains, 2440 meters (8000 ft) higher than *Apollo 11*. Using an electric car, Duke and Young made three trips lasting over 20 hours and gathered 97 kg (213 lb) of rock.

**Apollo 17** (7–19 December 1972)
Crew: Eugene A. Cernan, Harrison H. Schmitt, Ronald E. Evans.
Command Module: *America*; Lunar Module: *Challenger*.
*Challenger* landed in a valley west of the Sea of Serenity. Cernan and Schmitt covered a record 35 km (22 miles) in their lunar rover and gathered 115 kg (250 lb) of rock and soil. This was the last of the Apollo missions.

▲ *Edwin Aldrin, the second man on the Moon, photographed by the first: Neil Armstrong. Armstrong and the lunar module* Eagle *are reflected in Aldrin's visor.*

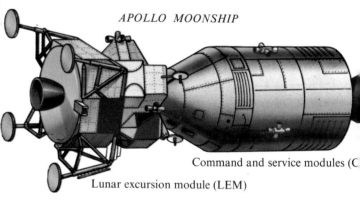

APOLLO MOONSHIP
Command and service modules (C
Lunar excursion module (LEM)

▲ *The crew sat in a cone-shaped command module joined to a fat tube. This service module provided power and other systems. A flimsy lunar module took two crewmen to the Moon.*

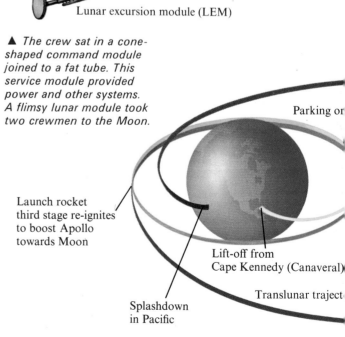

Parking or
Launch rocket third stage re-ignites to boost Apollo towards Moon
Lift-off from Cape Kennedy (Canaveral)
Translunar traject
Splashdown in Pacific

# Exploring the Moon

People began to explore the Moon with unmanned probes. The task was huge. The Earth—the launching pad—was spinning and moving in one direction. The Moon—the distant target—was moving in another. Then, too, each probe was pulled three ways: by gravity exerted by Earth, Moon and Sun. The target could not have been reached without many calculations worked out with the help of computers.

But success came quickly. In 1959 Russia's *Luna 2* hit the Moon, and *Luna 3* passed and photographed its hidden side. In the mid 1960s American *Ranger* probes took television pictures. Clearer pictures of the surface came in 1966 from *Luna 9* and America's *Surveyor 1*, which actually landed softly. Later *Surveyor* probes scooped up and studied Moon rock. Meanwhile American *Lunar Orbiter* probes photographed most of it.

By 1968 American technologists felt they knew enough to send men to the Moon. They also had the necessary tools, including the *Apollo* spacecraft and a landing vehicle. In December 1968 the *Apollo 8* mission took three astronauts ten times around the Moon. Two more missions followed. Then, in July 1969, *Apollo 11*'s lunar module landed. Its crew became the first two men to set foot on the Moon. This and later missions are listed on the page opposite.

▲ A scene from the Apollo 17 mission. The lunar rover stands in the foreground. Beyond, Harrison H. Schmitt works near a great boulder. In the distance is land lying between the Littrow Crater and Taurus Mountains. Because the Moon lacks air, the sky appears black.

▼ The lunar module from Apollo 12 on the Moon. This flimsy craft was made of light-weight aluminum alloy, partly sheathed in gold foil to reflect the burning rays of the Sun. It stood on four folding legs.

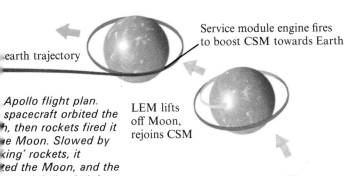

Apollo flight plan. spacecraft orbited the h, then rockets fired it e Moon. Slowed by king' rockets, it ted the Moon, and the r module landed. Later pper part joined the mand module, and was cast off. The service ule was later dropped. the small, scorched mand module came e.

Service module engine fires to boost CSM towards Earth

earth trajectory

LEM lifts off Moon, rejoins CSM

LEM separates from CSM, lands on Moon

Retrofire to place Apollo in lunar orbit

◀ The size of the Moon compared with Australia. If you could bring the two together in this way, you would notice that the distance across Australia is a little greater than the diameter of the Moon.

# A Dead World

Telescopes, lunar probes, and manned landings have told us much about the Moon. We know its gravity is low, it lacks air and water, its days are roasting hot, and its nights intensely cold. No plants or animals can survive on the Moon.

Looking at the Moon from Earth, you notice that some parts are dark and others pale. A telescope shows that the dark areas are fairly smooth. Astronomers call them "seas". But they are really bone-dry plains. Some are almost round, others have ragged edges. Both are floored with lava that welled up from inside the Moon perhaps 3500 million years ago.

The bright areas are highlands with huge mountain ranges and deep craters. Their rocks are even older than those of the lunar "seas", and largely made of rock chips stuck together.

Everywhere you look are craters. Some are tiny. Clavius, among the biggest, is 3650 meters (12,000 ft) deep and 233 kilometers (145 miles) across. Some of these craters are volcanic. But all the larger ones were gouged out when meteorites struck the surface of the Moon. From craters such as Kepler and Copernicus, white streaks stick out like the spokes of a wheel. These are probably made of glassy rock that splashed out as hot liquid when the craters were made.

So many meteorites and asteroids have hit the Moon that they have largely broken up its surface. Most Moon scenes show rocks and stones. In places the surface is metres deep in rock chips, dust and tiny glassy grains.

**Facts about the Moon**
Diameter: 3476 km (2160 miles), $\frac{1}{4}$ that of the Earth
Volume: $\frac{1}{49}$ that of the Earth
Mean density: 3·34 (density of water = 1)
Mass: $\frac{1}{81}$ that of the Earth
Gravity: $\frac{1}{6}$ that of the Earth
Free water: none
Atmosphere: none
Temperature: at noon 100°C; at midnight −150°C
Highest peaks: almost 6000 meters (20,000 feet) in the lunar Apennine Range
Largest crater: Orientale Basin, 965 km (600 miles) across
Escape velocity: 2·41 km/sec (1·5 miles/sec)
Turns on axis in: $27\frac{1}{3}$ days
Average distance from Earth: 384,402 km (238,856 miles)
Average orbiting speed: 3700 kph (2300 mph)
Orbits Earth: once in $27\frac{1}{3}$ days (sidereal month)
Synodic month (new Moon to new Moon): $29\frac{1}{2}$ days

### Inside the Moon

To find out about the Moon's interior, *Apollo* astronauts set up earthquake-detecting devices called seismometers. These showed that several thousand moonquakes happen every year. They seem to start far below the surface. Other studies hint that—like Earth—the Moon has a crust, a mantle and a core.

People once believed the Moon was torn out of the Earth. But its strange rocks suggest that the Moon had a separate beginning.

◀ *A strange patch of orange Moon soil, photographed in 1972 by Apollo 17 astronauts in the Taurus-Littrow area. Moon soil largely consists of dust, rock chips and tiny glass balls. Much of the soil is formed from the remains of meteorites and asteroids that have hit the Moon.*

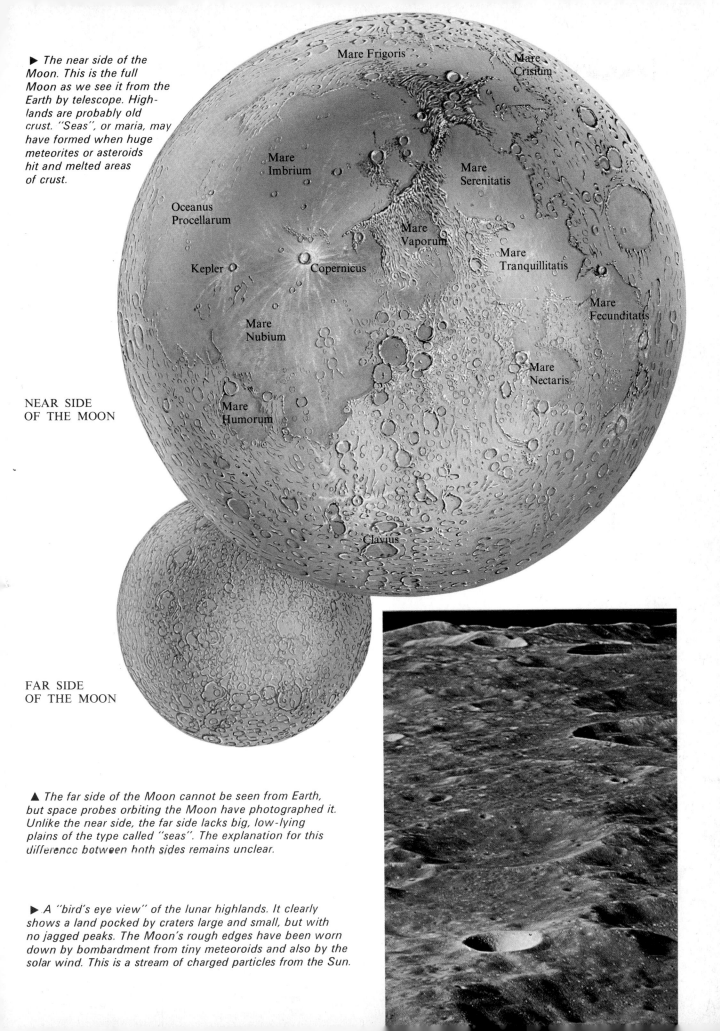

► The near side of the Moon. This is the full Moon as we see it from the Earth by telescope. Highlands are probably old crust. "Seas", or maria, may have formed when huge meteorites or asteroids hit and melted areas of crust.

NEAR SIDE OF THE MOON

FAR SIDE OF THE MOON

▲ The far side of the Moon cannot be seen from Earth, but space probes orbiting the Moon have photographed it. Unlike the near side, the far side lacks big, low-lying plains of the type called "seas". The explanation for this difference between both sides remains unclear.

► A "bird's eye view" of the lunar highlands. It clearly shows a land pocked by craters large and small, but with no jagged peaks. The Moon's rough edges have been worn down by bombardment from tiny meteoroids and also by the solar wind. This is a stream of charged particles from the Sun.

PHASES OF THE MOON

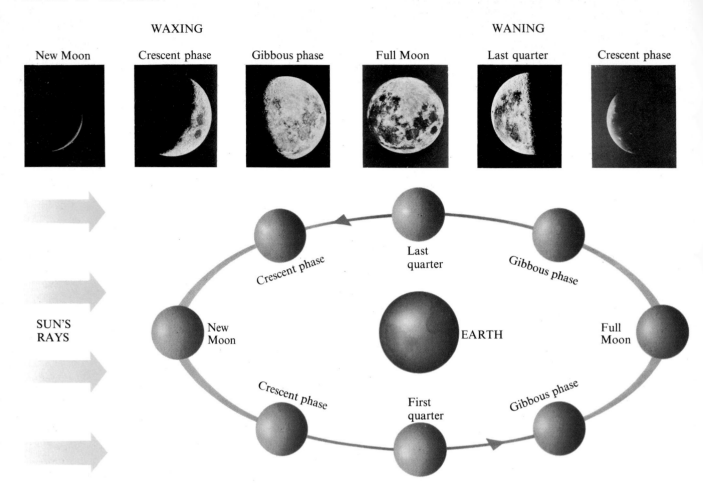

▲ The Sun's rays always shine upon the Moon. But the sunlit area we see grows and shrinks as the Moon revolves around the Earth creating the phases of the Moon.

# Moon and Earth

Some people hope that people might one day mine minerals upon the Moon. But many scientists believe the cost would be too high to make the task worthwhile. They see the Moon as a dead world too far away in space to help us.

Yet the Moon has always affected people—by its monthly path around the Earth, by helping to create the tides, and by casting a kind of shadow known as an eclipse.

## The Changing Moon

The Moon tours around the Earth once every 29 days. All that time, one half of the Moon is bathed in sunlight. But the angle that the Moon makes with the Earth and Sun is changing. This alters the amount of Moon that we can see from Earth. We call these changes *phases*. When the Moon lies between the Earth and Sun, the side of the Moon facing the Earth is dark. We call this the *new Moon* phase. We also use "new Moon" to describe the thin slice of Moon that shows up a day later. As the Moon moves on around the Earth it *waxes*, or seems to grow larger. The thin slice becomes a crescent. A week after new Moon the Moon resembles a half circle. This is its *first quarter*. A week later we can see the whole face at *full Moon*. This is the Moon's *second quarter*.

Then the Moon starts to *wane*. A week later comes the *last*, or *third*, *quarter*, another half circle. In its last phase the visible Moon is a crescent that thins to nothing. When the Moon appears between full and half circle in shape, astronomers say that it is *gibbous*.

Each time it circles the Earth, the Moon spins once upon its axis. This is why the same side of the Moon is always facing us.

## Moon and Tides

The Moon's gravity pulls only one-sixth as hard as the Earth's gravity. This is why men on the Moon weigh one-sixth as much as they weigh down on Earth.

But the Moon's pull is still strong enough to cause the oceans' tides. On the side of Earth facing the Moon, the Moon's pull draws ocean water up toward it. The solid Earth is also moved, but not so strongly. This pull draws it away from the water on the far side of the world. The results are high tides on opposite sides of the world. In between, the level of the sea drops. Here, low tides occur. As the Earth turns, the two high tides and two low tides sweep around the world and reach each ocean shore twice a day.

## Eclipses of the Sun and Moon

Sometimes the Moon's shadow falls upon part of the Earth. The shadow blots out part or all of the Sun. If all the Sun is hidden, the Earth becomes dark. Astronomers call such events *eclipses* of the Sun. They occur when the Moon lies between the Earth and Sun, so that a straight line would pass through all three. Sometimes the Sun casts Earth's shadow on the Moon. Astronomers call this event an eclipse of the Moon.

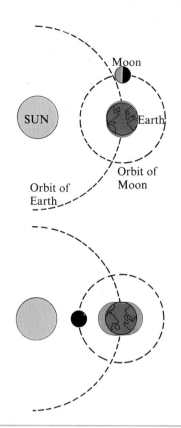

▶ Tides occur on Earth as the Moon's gravity pulls the oceans on the side of the Earth facing the Moon. The Earth is also slightly pulled towards the Moon and away from the waters that are farthest from the Moon. So the waters bulge on both sides of the Earth. As the Earth revolves, both of these high tides sweep around the world once each day. The Sun also affects the tides. It is bigger than the Moon but much farther away from the Earth. When the Sun and Moon pull at right angles, the Sun's gravity weakens the Moon's effect and produces neap tides. The tide then rises and falls less than at any other time (top diagram). When Sun and Moon pull together (bottom diagram), they produce spring tides.

## SOLAR AND LUNAR ECLIPSES

◀ The Moon as a dark ball during a total solar eclipse. The bright glare comes from the Sun's corona, or outer atmosphere. Normally you cannot see the corona because it is lost in the blinding glare from the Sun's surface, known as the photosphere. A total solar eclipse lasts for no more than $7\frac{1}{2}$ minutes at the most.

▼ How solar and lunar eclipses happen. In a solar eclipse, the Moon's shadow falls on the Earth. In a lunar eclipse, the Earth's shadow falls on the Moon. A dark shadow like each of the shadows shown is called an umbra. A partial shadow (not shown) surrounding the umbra is called a penumbra.

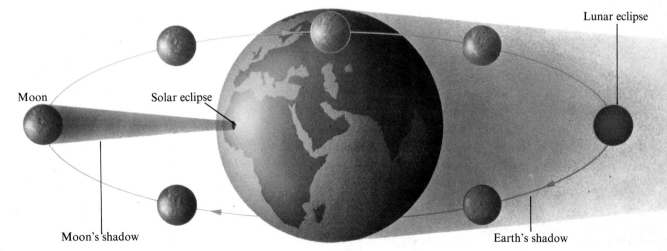

# Chapter Seven
# The Solar System

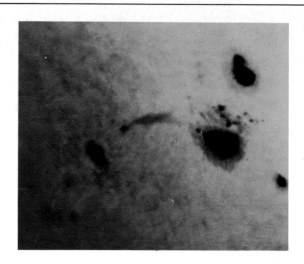

*Sunspots are marks like ink blots on the Sun's surface. Sunspots are some 2000°C below the temperature of the nearby surface. Ones larger than the Earth may last for months, others only for hours. The number of sunspots reaches a peak every 11 years.*

Our own Earth and other planets, as well as moons and small lumps of rock, hurtle around the star we call the Sun. The Sun and all these objects revolving around it make up the solar system.

The Sun itself is easily the largest, hottest object in the solar system. Its mass is 750 times that of all the objects whirling round it. Solar gravity prevents the moons and planets flying off in space.

Like most other stars, the Sun consists of gas. Nuclear reactions in its core change hydrogen to helium. This produces temperatures of 15,000,000 °C. Vast amounts of heat, light, and other kinds of energy shoot out through space as electromagnetic waves. Only the Sun's energy makes life possible on Earth.

**Facts about the Sun**
Diameter: 1,392,000 km (865,000 miles), 109 times greater than the Earth's
Volume: 1,303,600 times that of Earth
Mean density: 1·41 (density of water is 1)
Mass: 332,000 times that of Earth
Gravity: 28 times that of Earth
Temperature: Core 15,000,000°C; surface 6000°C
Energy consumed: 4,000,000 tons hydrogen/sec
Escape velocity: 618 km/sec (384 miles/sec)
Turns on axis in: 25·38 days
Distance from Earth: 150,000,000 km (93,000,000 miles)
Orbits galaxy once every 225 million years
Orbits at: 250 km/sec (155 miles/sec)
Sunlight reaches Earth in $8\frac{1}{3}$ minutes

▶ *Cutaway view of the Sun. Nuclear reactions at its core produce energy which is radiated from its surface. Here gases swirl, and burning tongues of gas (solar prominences) leap thousands of miles out into the atmosphere.*
▼ *Four photographs show a solar prominence rising.*

SUN    Mercury    Venus    Earth    Mars    Jupiter

# Family of Planets

▲ *After the Sun, the planets are the largest objects in the solar system. The bottom of the diagram shows how they compare in size with one another and the Sun. Straight lines lead from the planets to their orbits. These are shown as blue loops centered on a tiny orange ball (the Sun). The hazy blue belt between the orbits of Jupiter and Mars shows the orbits of asteroids, or minor planets. A long orange loop shows the orbit of a comet that swings far out into the solar system.*

Long ago astronomers noticed that certain "stars" moved around instead of seeming to keep still like all the rest. Astronomers later realized that this was because they were much nearer to Earth. Greeks gave these objects the name *planets*, meaning "wanderers". We now know, of course, that the planets are not stars at all. Unlike stars they do not produce their own light. They shine with light reflected from the Sun.

Nine planets, including Earth, orbit the Sun, and some are orbited by a moon or moons. The planets loop around the Sun at different distances. Their paths are not true circles, but are shaped like ellipses, or ovals. All of them orbit the Sun in the same direction. This is anti-clockwise.

Strangely, too, most of the planets orbit at almost the same plane, or level. These are two reasons why astronomers think that the planets formed from a great belt of dust and gas that once whirled around the Sun.

With no telescopes, early astronomers could see only the nearest planets to the Sun. Going outward from the Sun these are Mercury, Venus, Earth, Mars, Jupiter and Saturn. The three planets farthest out remained unknown for centuries. Astronomers discovered Uranus in 1781, Neptune in 1846, and Pluto in 1930.

Saturn      Uranus   Neptune   Pluto

## Groups of Planets

Astronomers group planets in various ways. Those nearer to the Sun than we are are called *inferior planets*. Those farther away are *superior planets*. Another grouping divides planets into Earth-like, or *terrestrial*, and *giant* planets. Mercury, Venus, Earth, Mars and Pluto are terrestrial planets. Each of them is small and dense with a hard surface, a thin atmosphere or none, and few or no moons.

Jupiter, Saturn, Uranus and Neptune are the giant planets. Each is larger and less dense than any Earth-like planet, and has a thicker atmosphere. Each giant planet also has at least two moons. If the smallest planet, Mercury, was pictured as a pinhead, the largest planet, Jupiter, would be as big as a golf ball. On the same scale, the Sun would be as big as a beach ball.

Planets vary in the force of gravity they exert. Mercury's gravity is far lower than the Earth's. Rather like astronauts on the Moon, visitors to little Mercury could leap about with ease, unless weighed down with heavy suits and boots. On the other hand, Jupiter has more than twice the force of gravity exerted by the Earth. If man could land upon this giant planet, his weight would seem to more than double, and it would take tremendous efforts to make even the smallest movements.

## Time and the Planets

The time a planet takes to orbit once around the Sun is that planet's year. Mercury's year is only 88 Earth-days long, but a year on Pluto lasts nearly 248 Earth-years. Planets spin as they orbit around the Sun. On any planet, night falls at each place as it turns from the Sun. But planets spin at different rates, so some have longer nights than others. On Venus the night is very long, because Venus takes 243 Earth-days to make one spin.

# Mercury

▼ *In 1974 the American space probe Mariner 10 took this photograph of Mercury. In many ways Mercury resembles our own Moon. Without an atmosphere to shield it, the planet has been struck by meteoroids that have punched craters in the bald, rocky surface.*

Mercury is the nearest planet to the Sun. It may also be the smallest true planet. Mercury is far smaller than the Earth. Its face is barren and rocky, with mountains and craters like those on the Moon. Many of these craters were made by meteoroids. Like the Moon, Mercury lacks an atmosphere to shield its surface from meteorites pulled in by its weak force of gravity.

Because Mercury lies so near the Sun, its surface gets hot enough by day to melt lead. Because it has no atmosphere to store heat, its nights are cold enough to cool oxygen gas until it turns into liquid oxygen.

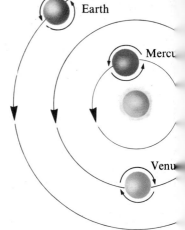

▲ *Mercury, Earth and most other planets spin in the same direction. But Venus revolves in the opposite direction.*

**Facts about Mercury**
Diameter: 4850 km (3015 miles), $\frac{1}{3}$ Earth's
Volume: $\frac{1}{16}$ that of the Earth
Density 5·4, about the same as Earth
Mass: $\frac{1}{18}$ that of the Earth
Gravity: about $\frac{1}{3}$ that of the Earth
Atmosphere: none
Temperature: day 400°C; night −200°C
Highest peaks: $2\frac{1}{2}$ km ($1\frac{1}{2}$ miles) above lowest land
Escape velocity: 4·2 km/sec (2·6 miles/sec)
Turns on axis in: 59 days
Distance from Sun: 58,000,000 km (36,000,000 miles)
Orbits Sun in: 87·97 days
Closest to Earth at: 45,000,000 km (28,000,000 miles)
Moons: none

▼ *An artist's impression of Mercury. Meteoroids made the craters. Cracks occurred as huge changes in the daily temperature made rocks swell and shrink.*

# Venus

◀ *Venus seen through a telescope a few days after passing between the Earth and the Sun. When Venus is directly between these two its face is hidden from us. Astronomers use the term "inferior conjunction" for this "new Moon phase" of Venus. The planet seems smallest but most fully lit when it lies beyond the Sun.*

Venus is the second planet from the Sun. It is not much smaller than the Earth, but it differs from the Earth in many ways. It has a shorter year but a far longer day, because it spins so slowly.

## A Shrouded World

Venus shows up brightly at dawn or sunset. People thus often call it the Morning Star or Evening Star. Venus shines brightly because it is closer to us than other planets are. Also it reflects sunlight from the thick clouds that hide its surface. The clouds probably consist of droplets of sulphuric acid.

Venus has a dense atmosphere, mainly of carbon dioxide gas. This atmosphere works like a greenhouse roof. It lets in the Sun's heat but stops much of it escaping. By day the planet's surface may be even hotter than that of Mercury. It stays warm by night too. The land on Venus may be mainly smooth, but radar pictures show some huge craters. Here and there may lie pools of molten metals.

**Facts about Venus**
Diameter: 12,140 km (7545 miles), similar to Earth's
Volume: about $\frac{9}{10}$ that of the Earth
Density: 5·2, much as that of Earth
Mass: about $\frac{4}{5}$ that of Earth
Gravity: nearly $\frac{9}{10}$ that of Earth
Atmosphere: mainly carbon dioxide
Temperature: up to 500°C
Highest peaks: unknown
Escape velocity: 10·3 km/sec (6·4 miles/sec)
Turns on axis in: 243 days
Distance from Sun: 108,000,000 km (67,000,000 miles)
Orbits Sun in: 224·7 days at 35 km/sec (21·7 miles/sec)
Closest to Earth at 42,000,000 km (26,000,000 miles)
Moons: none

*EXPLORING VENUS*
*A Soviet* Venera *capsule parachuting down to Venus. The instruments inside showed that the surface of Venus is only as dense as the Earth's skin of soil.*

▼ *Swirling clouds mask the surface of Venus. This photograph was taken by* Mariner 10 *in 1974 when the spacecraft was 720,000 kilometers (450,000 miles) away.*

# Earth in Space

Earth is the third planet from the Sun. Its surface waters and the gases making up its atmosphere help to keep Earth's surface comfortably warm, and damp. It is the only planet with known living things.

Like other planets, Earth spins. It completes one turn every 24 hours. Because it spins eastward, we see the Sun "rise" in the east and "sink" in the west. The Earth turns about its axis. This imaginary line passes through the Earth's center and its ends form the North Pole and South Pole.

The Earth is tilted as it orbits around the Sun. When Earth's northern half (the northern hemisphere) tilts toward the Sun, its days are long and nights are short. It is summer there. When it tilts away, days are short and nights are long. It is winter there. When days and nights are roughly equal, it is spring or fall. Thus the Earth's tilt produces the seasons.

Earth's hottest places are the tropics. Here the sun shines straight down at least once a year. In the cold polar regions, it fails to rise at all at least once a year.

▲ The setting Sun. Soon it will sink below the horizon, and night will have fallen. People watched the Sun rise and set for thousands of years without understanding what happens—that day and night sweep round the Earth as it revolves once every 24 hours in an eastward direction. Astronomers used to believe that Sun, Moon, and planets all circled the Earth, and that Earth was the center of the universe.

▶ Seasons occur on Earth because its axis is tilted by $23\frac{1}{2}$ degrees to the plane in which it orbits the Sun. The northern hemisphere leans farthest towards the Sun about 21 June. On that day the Sun shines directly overhead above the Tropic of Cancer. This is an imaginary line parallel with the equator. It is 23 degrees and 27 minutes north of the equator (about one-quarter the way from the equator to the North Pole). In the northern hemisphere it is now the longest day and the first day of summer. About 21 December the Sun shines directly over the Tropic of Capricorn in the southern hemisphere. In the northern hemisphere this is the shortest day and the first day of winter.

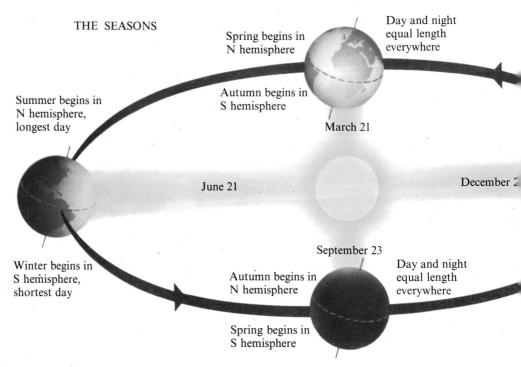

THE SEASONS

▶The Earth photographed from Apollo 11 in July 1969. Night has fallen on eastern Asia, but the Sun is shining on southwest Asia, southern Europe and all of Africa. Cloud covers parts of the land and much of the sea. Cloud "swirls" like the one west of Spain and Morocco are depressions—areas of low atmospheric pressure where warm tropical air meets cold polar air to give drizzle and showers. Such views of the Earth from space help weather experts to plot the paths of hurricanes and so to warn ships of dangerous storms.

Winter begins in N hemisphere, shortest day

Summer begins in S hemisphere, longest day

▲ Life near the Arctic Circle. The winters are bitterly cold and the nights so long that the Sun does not rise around 21 December. Around 21 June, however, the sun does not set.

**Facts about the Earth**
Diameter at equator: 12,756 km (7926 miles)
Circumference around poles: 40,007 km (24,859 miles)
Density: 5·52 (density of water is 1)
Mass: 5976 million million million tons
Area: 510,066,000 sq km (196,935,000 sq miles)
Surface: 71% water, 29% land
Temperature: lowest known −88·3°C; highest 57·7°C
Highest peak: Mt Everest 8848 m (29,028 ft)
Deepest ocean trench: 11,033 m (36,197 ft)
Escape velocity: 11·2 km/sec (7·0 miles/sec)
Turns on axis in: 23 hours, 56 min, 10 sec
Distance from Sun: 150,000,000 km (93,000,000 miles)
Orbits Sun in: 365 days, 6 hours, 9 min, 10 sec
Orbital velocity: 29·8 km/sec (18·5 miles/sec)
Moons: one

# Mars

Mars is the fourth planet out from the Sun. In some ways, it resembles Earth more than any other planet does. Like the Earth it has a solid rocky surface and an atmosphere. A day on Mars lasts only 41 minutes longer than a day on Earth. Mars is tilted at almost the same angle as the Earth, and so it has seasons too.

In other ways Mars is unlike Earth. It is only half as large as our own planet. Its year lasts nearly twice as long as ours. Then, too, it has two tiny moons: Deimos and Phobos.

Astronomers were able to find out quite a lot about Mars by watching it from Earth. The markings and colors on its surface show up quite well through telescopes. Because of its red color Mars has been nicknamed the "Red Planet". Some of Earth's deserts are reddish, so many astronomers supposed that most of Mars was also desert. But they noticed pale patches at the poles. These patches shrank in the Martian summer and grew in winter. Astronomers believed them to be polar ice caps, and the large, dark areas to be oceans and seas.

In 1877, the Italian astronomer Giovanni Schiaparelli said that he saw straight lines on Mars. He thought that they must be canals. If he was right, there might be life on Mars.

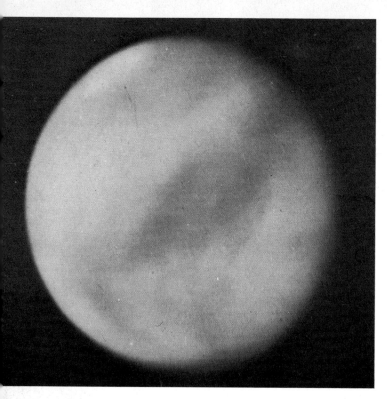

▲ Mars viewed from Earth by telescope. Until the last few years this was the only way of learning what kind of place the reddish planet really was. Telescopes showed that polar ice caps grew and shrank as seasons changed. They also showed strange markings that shifted with the seasons. Only space probes could solve these mysteries.

**Facts about Mars**
Diameter: 6790 km (4220 miles), about half Earth's
Volume: about $\frac{1}{7}$ that of the Earth
Density: 3·95, about $\frac{2}{3}$ that of Earth
Mass: about $\frac{1}{9}$ that of the Earth
Gravity: more than $\frac{1}{3}$ that of the Earth
Atmosphere: thin, mainly carbon dioxide
Temperature: below −180°C to just above 0°C
Highest peak: Nix Olympica, 30 km (19 miles) high
Escape velocity: 5·0 km/sec (3·1 miles/sec)
Turns on axis in: 24 hours 37 min
Distance from Sun: 228,000,000 km (142,000,000 miles)
Orbits Sun in: 687 days
Moons: 2 (Phobos and Deimos)

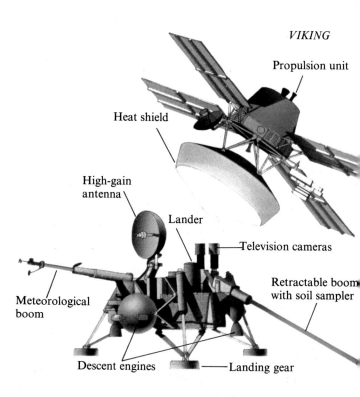

▶ Much of what we now know about Mars is information beamed to Earth from unmanned space probes, especially Viking spacecraft—the type shown here. In summer 1976 the United States set two Vikings in orbit around the distant planet. Both released capsules which touched down softly on the surface. The space probes sent back pictures of the planet and facts about its soil and weather.

## Messages From Mars

Since the 1960s began, the Soviet Union and the United States have sent space probes to Mars. In 1976 the first American *Viking* spacecraft actually landed on the planet after a year-long journey of 800 million kilometers (500 million miles). Soon it was radoing pictures back to Earth. Traveling at the speed of light they took 18 minutes to span the gulf of space. These pictures showed that some old ideas about Mars were right, but others wrong.

We now know that Mars indeed looks much like parts of our Arizona Desert. The surface is made up of rocks and sand. These surface rocks seem rich in iron. Scientists have also discovered aluminum, calcium, and other elements.

But there is no trace of oceans. Dark markings on the surface appear and disappear as dust storms whirl across it. The ice caps that grow and shrink are largely made of carbon dioxide gas, so cold that it is frozen solid. Here and there water has gouged valleys through the land. But there are no canals, and there is now no sign of liquid water on the surface.

Careful tests have found no certain trace of life. Bitter cold and lack of air would make life difficult. True, there is enough atmosphere for winds to blow. But the atmosphere is probably one-tenth as dense as it once was. Its pressure is a mere one-hundredth of that exerted by the atmosphere on Earth.

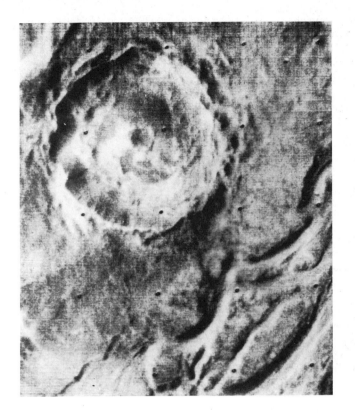

▲ *Mariner 9 took this photograph of Mars in 1972 while orbiting the planet. It shows a crater gouged out by a falling meteoroid. Below, you can see part of a snaky channel, perhaps carved out long ago by running water.*

▼ *After soft-landing on Mars in 1976, Viking 1 took this picture of the nearby landscape. It shows rocks strewn among what seem to be dunes made up of reddish sand. The scene is similar to some deserts here on Earth.*

# The Outer Planets

**Facts about Jupiter**
Diameter: 142,600 km (88,600 miles), 11 times Earth's
Volume: 1316 times that of Earth
Density: 1·34, about $\frac{1}{4}$ that of Earth
Mass: 317·8 times that of Earth
Gravity: 2·64 that of Earth
Escape velocity: 61 km/sec (38 miles/sec)
Turns on axis in: 9 hours 50 min
Distance from Sun: 778,000,000 km (484,000,000 miles)
Orbits Sun in: 11·9 years
Moons: 14

**Facts about Saturn**
Diameter: 120,200 km (74,700 miles), 9·4 times Earth's
Volume: 755 times Earth's
Density: $\frac{1}{8}$ that of Earth
Mass: 95·2 times that of Earth
Gravity: 1·2 times that of Earth
Escape velocity: 37 km/sec (23 miles/sec)
Turns on axis in: 10 hours 14 min
Distance from Sun: 1,427,000,000 km (887,000,000 miles)
Orbits Sun in 29·5 years
Moons: 10

**Facts about Uranus**
Diameter: 49,000 km (30,500 miles), 3·8 times Earth's
Volume: 52 times Earth's
Density: 1·58, about $\frac{1}{3}$ Earth's
Mass ('weight'): 14·5 Earth's
Gravity: 1·1 Earth's
Escape velocity: 22 km/sec (14 miles/sec)
Turns on axis in: about 24 hours
Distance from Sun: 2,870,000,000 km (1,783,000,000 miles)
Orbits Sun in: 84 years
Moons: 5

**Facts about Neptune**
Diameter: 50,200 km (31,200 miles), 3·9 times Earth's
Volume: 44 times Earth's
Density: 2·30, less than half Earth's
Mass: 17·2 Earth's
Gravity: 1·4 Earth's
Escape velocity: 25 km/sec (16 miles/sec)
Turns on axis in: about 22 hours
Distance from Sun: 4,497,000,000 km (2,794,000,000 miles)
Orbits Sun in: 164·8 years
Moons: 2

**Facts about Pluto**
Diameter: 6400 km (3980 miles), about $\frac{1}{2}$ Earth's (?)
Mass: 0·17 that of Earth
Turns on axis in: 153 hours (?)
Average distance from Sun: 5,900,000,000 km (3,670,000,000 miles)
Orbits Sun in: 247·7 years
Velocity in orbit: 4·7 km/sec (2·9 miles/sec)
Moons: 1

◄ Uranus and the five moons that circle it. None is more than 1000 km (620 miles) across. More facts about the farther planets are still being discovered. In 1977, for instance, astronomers found that Uranus has rings rather like those of Saturn.

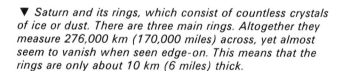

▼ Saturn and its rings, which consist of countless crystals of ice or dust. There are three main rings. Altogether they measure 276,000 km (170,000 miles) across, yet almost seem to vanish when seen edge-on. This means that the rings are only about 10 km (6 miles) thick.

Beyond the Earth and Mars are the five planets farthest from the Sun. The first four are giants far larger than the Earth. They are Jupiter, Saturn, Uranus and Neptune. Compared with them, Pluto is a dwarf. This fifth outer planet is smaller than the Earth.

Jupiter is twice as massive as the combined sizes of all other planets. It spins faster and has more moons than any of the rest. Beneath its cloudy surface this strange giant may be mainly made of liquid hydrogen.

Saturn is the second largest planet. Yet it is mainly made of three gases: hydrogen, helium and methane. Saturn is a featherweight giant that would float in water. Flat rings encircle its equator. They make this planet the most beautiful of all to look at through a telescope.

Uranus is about four times as far across as the Earth. Like Jupiter and Saturn, it is mainly made of gases. Uranus is intensely cold. A surprising fact was discovered as recently as 1977. Uranus has a ring system like Saturn's only very much fainter and smaller.

Neptune is very like Uranus in its size, make-up, and its bitter cold. But it has only two moons compared with the five of Uranus.

Pluto is the remotest planet. It is nearly 40 times farther from the Sun than we are. Pluto may be even smaller than Mercury.

▶ An arrow pinpoints Pluto, the remotest, and second smallest planet. Seen from Earth, Pluto is so tiny that no one identified it until 1930. Even the vast 200-inch (508-cm) Hale telescope reveals it only as a point of light that moves against the stars far out in space beyond it. We know less about this distant planet than we do about most of the rest.

▶ Neptune and (arrowed) its larger moon, Triton. This is about 3800 km (2360 miles) across—not much smaller than the planet Mercury. It orbits close to Neptune. The planet's other moon is far smaller. Nereid measures only 240 km (150 miles) across. It orbits Neptune far out in space. If shown on this scale Nereid would appear off the edge of the page.

▶ The giant planet Jupiter seen through a telescope. Clouds hide its surface. The planet spins so fast that it pulls the clouds into dark and bright stripes. Astronomers think the bright stripes are areas of rising atmosphere, and the dark stripes are troughs where the atmosphere descends. Beneath the clouds, they think, lies hot, liquid hydrogen. A strange surface feature is the Great Red Spot, seen in Jupiter's southern hemisphere. It measures 40,000 km (25,000 miles) long. American space probe Pioneer 10 found that this spot seems to be a storm that has swirled for centuries. The small dark spot is a shadow cast by one of the 13 moons that orbit Jupiter.

▲ About 25,000 BC an iron-nickel meteorite up to 79 meters (260 ft) across, and weighing 2,000,000 tons, gouged this hole in northern Arizona. The Coon Butte Crater is the world's largest meteorite crater.

# Junk in Space

Besides the planets and their moons, countless smaller objects whirl around our Sun. These are meteoroids, asteroids and comets.

**Meteoroids, Meteors and Meteorites**

Meteoroids are mainly tiny. Many are no larger than a grain of sand. Such tiny bodies are called micrometeoroids.

As they speed through space, many meteoroids are captured by Earth's gravity. They then plunge downward at up to 65 kilometers (40 miles) per second. When they strike the upper atmosphere most burn up and disappear as gases. People see them as brief streaks that dart in the night sky, and call them shooting stars or falling stars. The scientific name is *meteors*. Sometimes the Earth travels through a cloud of meteoroids. Then the result is a *meteor shower*. The Leonids and Perseids are meteor showers that reappear.

Some meteoroids are big enough to hit the Earth without burning up. Astronomers call them *meteorites*. Meteorites are made of iron or stone, or both. Most are small. But the Hoba Meteorite found in South-West Africa weighs almost 60

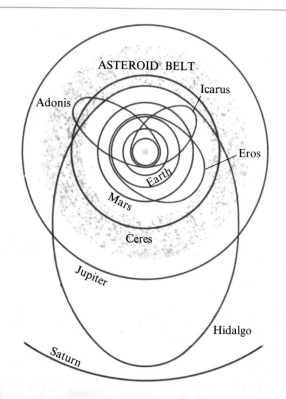

◄ The short streak on this photograph is one of three detected when astronomers examined the whole photograph. Each streak betrayed an asteroid, or minor planet, seen moving against the background of the stars. Left: most asteroids orbit the Sun in the so-called asteroid belt. This lies in a "gap" between the orbits of Jupiter and Mars. But some asteroids swing closer to the Sun and also farther out. This diagram shows the orbits of four wandering asteroids (Icarus, Adonis, Eros, and Hidalgo) and that of Ceres, the largest asteroid of all.

## THE PATHS OF THREE COMETS

The diagram over-printed on this page compares the orbits of three comets. Encke only takes 3·3 years to complete one orbit. It is called a short-period comet. Halley's Comet is a medium-period comet. It takes 76 years to swing close to Pluto's orbit and back. Kohoutek is a long-period comet. It needs at least 75,000 years to track far beyond Pluto and return.

tons. It is the largest ever found. A big meteorite may blast a huge hole in the ground, like a bomb crater. But most meteorites leave little trace. More than a million tons of them may fall on Earth as dust each year.

### Asteroids

Two centuries ago astronomers puzzled over the huge gap between Mars and Jupiter. Calculations suggested that this gap should hold a planet. Since then, observations have found not one but many thousand objects in this part of the solar system. Astronomers call them minor planets or *asteroids*. Asteroids are chunks of rock. Only twelve are more than 250 kilometers (155 miles) across. All asteroids put together would make a world a tiny fraction of the size of Earth.

### Comets

Sometimes "snowballs" with glowing tails shine in the sky night after night. They are comets. Comets probably consist of dust, ice and freezing gases. Comets tour the Sun in long orbits. As a comet nears the Sun, the Sun's heat melts the comet's ice and sets it glowing.

Some comets take far longer to reappear than others. Encke's Comet reappears once every 3·3 years. Halley's Comet comes round every 76 years. These *short-term* and *medium-term comets* do not pass beyond the planets. But Kohoutek takes 75,000 years to complete one orbit. Such *long-term comets* zoom far beyond the outer planets.

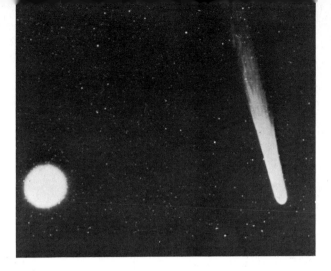

▲ A long tail flares from Halley's Comet in this photograph, taken in 1910. It was the comet's 29th known appearance. The bright object is Venus.

▲ Comet Humason was the fifth comet discovered in 1961. It is large, with a twisted tail. It will not reappear for nearly 2900 years.

▲ Comet Ikeya-Seki, 1965, was one of the brightest to be found this century. In places it even showed up plainly by day. It orbits once in 880 years or so.

# Chapter Eight
# Outer Space

▲ *Overhead and side views of the Milky Way as it would look if seen from outer space. They show that our own galaxy is a vast, flat, spinning spiral made up of myriads of stars. The red arrow marks our Sun's position, 30,000 light-years from the center. Many thousands of such spirals are among the 10,000 million galaxies in the universe.*

▶ *Some regions of our galaxy have more stars than others. A star cloud—a dense patch of stars—shines palely in this part of the Milky Way, seen edge-on from Earth. The pale line is the track of an artificial satellite.*

Beyond our Sun and its family of planets, an unaided human eye can see about 6000 stars. Telescopes show in fact that there are countless millions of stars strewn through space.

At least 100,000 million stars make up the great group of stars, or galaxy, to which our Sun belongs. This galaxy is called the Milky Way. Inside the Milky Way, the nearest star to the Sun is more than $4\frac{1}{4}$ light-years away. One light-year is the distance that light travels in one year. This is 9·5 million million kilometers (5·9 million million miles). The entire Milky Way is probably 100,000 light-years from side to side, and up to 20,000 light-years thick. It spins, and speeds through space at 2·1 million kilometers (1·4 million miles) per hour.

In and beyond the Milky Way lie millions of stars much like the Sun. Some of these glowing balls of gas are far larger than the Sun. One of these red giants with its center where the Sun is would actually engulf the solar system. Then there are suns much smaller than our own. White dwarfs are stars as tiny as Earth but so dense that a thimbleful of their material weighs tons.

There are stars that glow dimly, and stars that blaze more brightly than 100,000 suns. Certain stars send out light or other kinds of energy in pulses. Some stars break up with giant explosions. We are now seeing star groups far out across the universe by light that left them 10,000 million years ago. For all we know such stars and many others dimmed and fizzled out before our Sun began to shine.

Astronomers have learnt all this and much more thanks to telescopes and other special instruments. They can work out what a distant star is made of, how hot it is, how fast it spins, and other facts about it.

# Stars

Like a person, a star is born, matures, ages and dies. It starts life in a cloud of gas and dust known as a *nebula*. In time, gravity pulls some of the gas and dust close together to form a giant ball. As the substances inside the ball press in toward its center, the pressure there becomes intense. This makes the center very hot. Its temperature climbs yet higher as gravity pulls the gas and dust still farther inward. The shrinking ball glows even hotter.

Hydrogen gas starts turning into helium gas when it is hotter than 10 million degrees Centigrade. This nuclear reaction releases huge quantities of energy.

The ball is now a star. Once a big new star has settled down, it glows brilliantly blue or white for only a few million years. But a smaller star may last longer because its nuclear reactions tend to happen at a slower rate. Thus a star like our Sun may glow steadily yellow or red for thousands of millions of years. Some dwarfs may take 20 million million years to use up all their hydrogen.

**How Stars Grow Old**

When a star has used up all its hydrogen, fresh reactions happen. The center heats up, and the

*Most stars shine with a steady light. But seen from Earth the stars seem to twinkle. This is because their light has traveled through the atmosphere to reach us. Movement of different layers in the atmosphere distorts the light and gives a shimmering effect.*

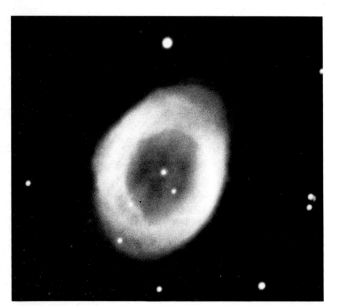

▼ *The Ring nebula in the constellation of stars called Lyra. The halo is an expanding shell of gas some 8 light-months across and 20,000 light-years from Earth. The gas probably spread outward when a central star exploded into a nova. This type of feature is a planetary nebula.*

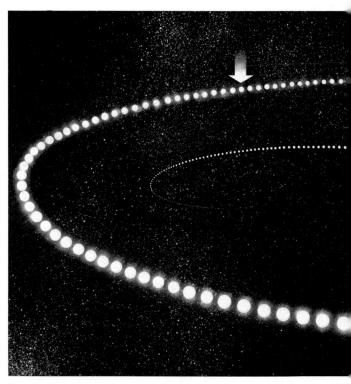

star swells until it is a giant or supergiant far larger than our Sun. Most stars then simply shrink to form white dwarfs—those faded stars no larger than the Earth.

Not all stars end so tamely. Big stars may suddenly explode and briefly shine as brightly as 100 million suns. This change can take a mere few days or even hours. The sudden brightness of such stars once led astronomers to think that new stars had appeared. They called these exploding giants *supernovae* (*nova* means "new"). Supernovae are extremely rare. The last one to show up in the Milky Way occurred nearly 400 years ago.

Once a supernova has exploded it leaves behind only expanding clouds of gas and a small dense object called a *stellar remnant*.

Our own Sun is already about 5000 million years old. It will probably shine on for 5000 million years more. Then it will bulge into a red giant big enough to swallow Mercury, Venus, Earth and Mars. Next it will shrink into a dense, white dwarf. Lastly it will cool and fade into the darkness of space.

## Stars that Change

Most stars that we see shine steadily. But some seem brighter at certain times than at others. They are called *variable stars*. *True variable* stars vary in brightness with changes in processes happening inside them. For instance the giant

▲ The brilliant Orion nebula probably consists of thinly spread hydrogen and other gases, and graphite dust. Hot stars inside this giant cloud set its gases glowing. The nebula is 1500 light-years distant and about 16 across.

red star Mira Ceti varies from least to most bright and back again every 11 months. Astronomers call Mira Ceti a *long-period variable*. Some of the so-called Cepheids are *short-period* stars. They vary over less than one day. The supergiant star Betelgeuse grows dim and bright irregularly. It is an *irregular variable*.

*Eclipsing variables* are stars that only *seem* to vary in brightness. They come in pairs. From time to time each star in one pair hides the other, and so affects the light we get from it.

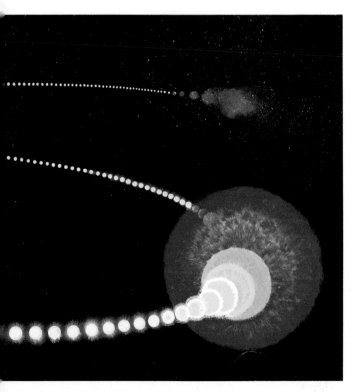

◀ An artist's idea of how a sun is born, grows old and dies. The sun forms when a cloud of gas condenses. Over thousands of millions of years, the star grows larger. Suddenly, it expands into a red giant. After this, the star begins to shrink and fade. In time it dwindles into a white dwarf and vanishes into the blackness of space. The arrow shows how far along this path our Sun is.

# Groups of Stars

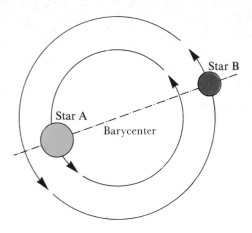

▲ A binary star system. Its stars are shown in brown and blue. Arrowed rings mark their orbits. The "blue" star is bigger than the "brown" star, but both revolve around the same center of gravity, called barycenter.

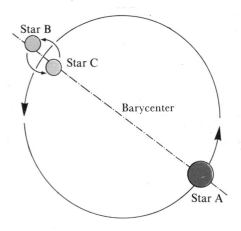

▲ A triple-star system. In this system the small stars B and C are close together and orbit a common center of gravity. Seen from Earth, B and C might be mistaken for a single star, so the system appears to be binary.

Most stars form groups of one kind or another. Nearly half of all stars belong to a *binary system*. In a binary system, two stars revolve around the same center of gravity. Telescopes have revealed more than 70,000 such binary systems.

Each member of a pair may be so near the other that astronomers have only found them with the aid of special instruments called *spectroscopes*. In some pairs, one partner sometimes hides the other from the Earth. The result is an eclipsing binary system. Its stars are the same as the eclipsing variables mentioned on page 81.

Nearly one-quarter of all stars belong to a *multiple system*—a system where three stars or more are linked by gravity. Close study shows that some "binary" systems really contain three stars, or even larger numbers.

## Larger Groups

Whole groups of stars move through space together. These *star clusters* include the Hyades and Pleiades in the constellation Taurus. Each of these clusters is about 15 light-years across and holds more than 100 stars.

Looser groups called *associations* hold young, scattered stars. One association may measure 1000 light-years across. Old stars occur largely in *globular clusters*. One which is 15 light-years across may have 10 million stars.

◄ A few of the stars in the open cluster called the Pleiades or the Seven Sisters. More than 100 of these hot blue stars can be seen glowing in the constellation Taurus. Dust clouds surround the chief stars with a bright haze. These stars were formed only a few million years ago. The whole group is about 15 light-years across and 400 light-years distant.

# Galaxies

Men have called galaxies "star islands in space". Galaxies are among the biggest groups of stars there are. One may hold about 100,000 million stars. Almost all have some sort of regular shape. Most are either spirals or ellipses. In a *spiral galaxy* clustered stars make up arms that spiral outward from the center. In an *elliptical galaxy* the stars form a shape like a partly tilted saucer seen from one side.

Galaxies themselves tend to be grouped in giant clusters. Our own Milky Way may belong to one such cluster—the local group—several hundred million light-years across. The universe may hold no fewer than 10,000 million galaxies. Some are so far off that their light takes thousands of millions of years to reach us.

▲ *The Large Magellanic Cloud is our nearest galaxy. It is 170,000 light-years away and 23,000 light-years across. It has the equivalent of 1,500,000,000 suns like ours.*

▼ *The two galaxies shown here represent different kinds of barred spiral galaxies. In a barred spiral galaxy, the arms start from the ends of a bar lying across the nucleus or center.*

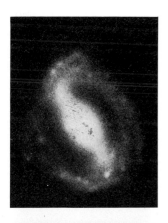

*Although 2·2 million light-years away, the Andromeda galaxy is one of the Milky Way's near neighbors. Its stars and nebulae measure 200,000 light-years across.*

# Star-Gazing

Taurus, the Bull

On a clear night you find that many bright stars make up patterns called *constellations*. They are not true star groups like the galaxies. But they are useful "landmarks" for finding individual stars. There are more than 80 constellations. Some reminded ancient astronomers of heroes, gods or beasts. This is how many constellations got their old Latin names.

Scorpio, the Scorpion

**CONSTELLATIONS OF THE NORTHERN HEMISPHERE**

1 Equuleus, Colt
2 Delphinus, Dolphin
3 Pegasus, Flying Horse
4 Pisces, Fishes
5 Cetus, Sea Monster
6 Aries, Ram
7 Triangulum, Triangle
8 Andromeda, Chained Maiden
9 Lacerta, Lizard
10 Cygnus, Swan
11 Sagitta, Arrow
12 Aquila, Eagle
13 Lyra, Lyre
14 Cepheus, King
15 Cassiopeia, Lady in Chair
16 Perseus, Champion
17 Camelopardus, Giraffe
18 Auriga, Charioteer
19 Taurus, Bull
20 Orion, Hunter
21 Lynx, Lynx

[22 Polaris, Pole Star]
23 Ursa Minor, Little Bear
24 Draco, Dragon
25 Hercules, Kneeling Giant
26 Ophiuchus, Serpent-Bearer
27 Serpens, Serpent
28 Corona Borealis, Northern Crown
29 Boötes, Herdsman
30 Ursa Major, Great Bear
31 Gemini, Twins
32 Cancer, Crab
33 Canis Minor, Little Dog
34 Hydra, Sea Serpent
35 Leo, Lion
36 Leo Minor, Little Lion
37 Canes Venatici, Hunting Dogs
38 Coma Berenices, Berenice's Hair
39 Virgo, Virgin

Ursa Major, the Great Bear

Cygnus, the Swan

Cassiopeia, the Lady in the Chair

Orion, the Hunter

Cancer,
the Crab

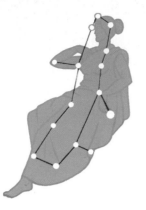
Virgo,
the Virgin

Centaurus,
the Centaur

Sagittarius,
the Archer

**CONSTELLATIONS OF THE SOUTHERN HEMISPHERE**

1 Cetus, Sea Monster
2 Sculptor, Sculptor
3 Aquarius,
  Water-Bearer
4 Piscis Austrinus,
  Southern Fish
5 Capricornus, Sea Goat
6 Grus, Crane
7 Phoenix, Phoenix
8 Fornax, Furnace
9 Eridanus,
  River Eridanus
10 Hydrus, Little Snake
11 Tucana, Toucan
12 Indus, Indian
13 Sagittarius, Archer
14 Aquila, Eagle
15 Corona Australis,
   Southern Crown
16 Pava, Peacock
17 Octans, Octant
18 Dorado, Swordfish
19 Pictor, Painter's Easel
20 Columba, Dove
21 Lepus, Hare
22 Orion, Hunter

23 Monoceros, Unicorn
24 Canis Major, Great Dog
25 Puppis, Poop
26 Carina, Keel
27 Volans, Flying Fish
28 Chamaeleon,
   Chameleon
29 Apus, Bird of Paradise
30 Triangulum Australe,
   Southern Triangle
31 Ara, Altar
32 Scorpio, Scorpion
33 Serpens, Serpent
34 Ophiuchus,
   Serpent-Bearer
35 Lupus, Wolf
36 Centaurus, Centaur
37 Crux, Southern Cross
38 Musca, Fly
39 Vela, Sails
40 Pyxis, Compass Box
41 Hydra, Sea Serpent
42 Sextans, Sextant
43 Crater, Cup
44 Corvus, Crow
45 Libra, Scales
46 Virgo, Virgin

Canis Major,
the Great Dog

*THE AURORAS*
*Star-gazing at night can reveal more than light from the stars. In the far north and south of the Earth, charged solar particles, striking the upper atmosphere, light up the night sky with glowing curtains, arcs and rays. These are the auroras, or Northern and Southern Lights.*

Capricornus,
the Sea Goat

# Space Mysteries

This century, astronomers have learnt much about the countless stars that twinkle out in space. But much remains to be discovered. Some of the biggest problems involve those strange objects known as quasars, pulsars and black holes.

In 1960 an American radio telescope found a faint star-like object that sent out radio "noise". Soon, astronomers found others. They called these star-like (quasi-stellar) radio sources *quasars*. Quasars appear to be small and bluish. They vary in brightness. In 1963 an American discovered a quasar a mere one ten-thousandth as big as an ordinary galaxy, yet 200 times as bright. Others seem to be retreating almost at the speed of light. Some experts think that quasars are caused by the collapse of giant galaxies. But no one can be sure.

In 1967 a British radio telescope detected radio waves pulsing out from somewhere deep in space. Astronomers invented the name *pulsars* for the objects sending out these pulses. Since 1967 many

## NEW DISCOVERIES

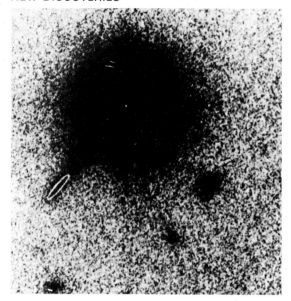

▲ This negative photograph shows the quasar 3C-273. It resembles a blue star with a projecting jet. The places from which it emits radio "noise" are shown by the white dot at the center and a white oval on the jet. Only 10 light-years across, it seems to be 2000 million light-years away, yet it is 200 times brighter than an average galaxy would be.

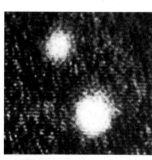

◀ Two photographs of the Crab pulsar, taken a fraction of a second apart. The larger star in the upper photograph is the pulsar emitting light. In the lower photograph it has "switched off" and become invisible. Besides light, it emits radio pulses with 100 times less energy, and X-ray pulses with 100 times more energy. The pulsating star that produces all this energy may be a spinning neutron star 30 km across.

Black holes may be even smaller and denser than the pulsars. Astronomers believe they represent the remains of big stars that have collapsed in on themselves until their density and gravity are too great for radiation to escape. Astronomers think they can detect black holes from the X-rays given off when gas clouds are heated as they are sucked into the "holes" from stars in nearby parts of space.

▶ An Unidentified Flying Object. Hundreds of such sightings are reported every year. Many turn out to be aircraft, or to have other rational explanations. Many remain unsolved mysteries. Some people think that they are signs of beings from another planet.

▼ What does space hold for the future? The US space agency, NASA, think that people will live in space in the future. This is an artist's impression of one of NASA's schemes. It is a space station, shaped like a wheel, and using the Sun's energy to grow plants. Some parts of the wheel would be used for agriculture, other parts for industry. The spinning wheel would produce an outward force which would seem similar to the Earth's gravity.

pulsars have been found. Pulsars give off huge quantities of energy yet may be no more than 30 kilometers (20 miles) across. Astronomers suggest they may be spinning neutron stars. These would consist of closely packed neutrons—particles making up the heavy core of atoms.

*Black holes* are smaller and denser even than the pulsars. Black holes may be big stars that collapsed in on themselves, so that no light or other form of radiation can escape.

## Visitors from Space?

Only astronomers with powerful telescopes can glimpse quasars, pulsars and the like. But many ordinary people claim to have seen even stranger objects from deep space. They say that they have seen mysterious lights and shapes in the sky. They argue that some of these UFOs (Unidentified Flying Objects) are spaceships from other worlds. But is this possible?

Astronomers believe there must be millions of planets in the universe. They think that some probably hold beings as intelligent as man. But such worlds would seem to be too far off to send us space explorers, unless they have developed new methods of travel. We may never know.

# FACT INDEX

This Fact Index contains some entries followed by a page number. These refer you to a page in this book where there is an article or information on that entry. Other entries supply information only. Page numbers in **bold type** refer to illustrations.

## A

**Abyssal plain** 16–17
**Aconcagua**, the highest volcano 21
**Adonis**, asteroid **76**
**Air** 14, 15, 27, 30, 47, 54, 55
**Aldrin, Edwin**, the second man on the Moon 55, **58**
**Alloy** 40
**Alps** 24, 25
**Aluminum** 40, 73
**Amazon, River**, largest river in the world 28
**Amphibian** 32
**Andes** 24, 25
**Andromeda galaxy 83**
**Animals** 30, 33, **36–37**, 38, 46, 47; extinct animals 37
**Animal realms 36–37**
**Antarctica** 24, 28
**Antarctic Ocean** 16
**Anteater 36**, 37
**Anticline** is an upfold in layers of rocks. **24**
**Anticyclone** is an area of high air pressure. Anticyclones tend to bring settled weather. See also *cyclone*.
**Apollo** (project) **51**, 55, **55**, 58–59, **58–59**, 60, **60**, 71
**Arctic Circle** 71
**Arctic Ocean** 16
**Aristarchus**, early astronomer 49
**Armstrong, Neil**, first man on the Moon 55, 58, **58**
**Artificial satellites** are man-made *moons*. 51, **52–53**, 55, 56
**Ash** (volcanic) 20, **21**
**Aso, Mount**, the largest volcanic crater 21
**Association** (of stars) 82
**Asteroids** are tiny planets. 10, 60, 67, **76**, 77
**Astrologers** are people who try to foretell future events by studying the stars' positions. 48

▼ *An anticline in Iran*

**Astronaut** 54–55, 59, 60
**Astronomers** are scientists who study objects outside Earth. 10, 48, 60, 66
**Atlantic Ocean** 16–17
**Atmosphere** Gases above the surface of a star or planet make up its atmosphere. 13, 14, **14–15**, 54, 70; of planets 67, 68, 69, 72, 73, **75**
**Atmospheric pressure** 54
**Atoms** are the smallest *particles* of matter to behave chemically in a special way. 14, 15, 43
**Aurora** 15, **85**
**Australasian realm 36**, 37, **37**
**Avalanche** 22
**Axis** The Earth spins around a line called its axis. 70; moon's axis 62

## B

**Bacteria** 30
**Barycenter** is the center of balance. **82**
**Basalt** 12, **20**
**Beach** 27
**Bessel, Friedrich**, famous astronomer 49
**Betelgeuse**, supergiant star 81
**Big Bang theory 8**, 10
**Binary star system 82**
**Biosphere** 30
**Birds** 30, **30**, 32, 33, 37
**Bison** 36
**Black holes** may be collapsed stars that let no light escape. 86, 87
**Blast furnace** 40, **41**
**Block mountains** 25
**Bondi, Hermann** 10
**Brahe, Tycho**, famous astronomer 49
**Bricks** 40

## C

**Cactus** 35
**Calcium** (on Mars) 73
**Cambrian period** 32
**Cancer**, the Crab (constellation of stars) **83**
**Canis Major**, the Great Dog (constellation of stars) **83**
**Capricornus**, the Sea Goat (constellation of stars) **83**
**Capsule** (space) 54–55
**Carbon dioxide** 26, 27; on Mars 73
**Carboniferous period** 32, **32**, 42
**Cars** 41, 44, 46
**Cassiopeia**, the Lady in the Chair (constellation of stars) **82**
**Caucasian people** 38

**Cave** 27, 30
**Cayley Plains** 58
**Cement** 40
**Cenozoic era** 32, **32**
**Centaurus**, the Centaur, (constellation of stars) **83**
**Cepheid** 81
**Ceres**, asteroid **76**
**Chalk** 27
**Chemicals** 26, 27, 30, 33, 47
**Chemical weathering** 27
**Chicago** 44
**Cirrus cloud** is thin, high, patchy cloud made of ice crystals. **15**
**City** 44, 45, 47
**Clavius**, crater on the Moon 60, **61**
**Clay** 27, 40
**Cliff 26**, 27
**Climate** 14, **34–35**, 38
**Cloud** consists of tiny water droplets or crystals of ice. It forms as water vapor cools in air. 11, 14, 17, 53, 71; on Jupiter **75**, on Venus 69
**Coal** 42, **42**, 43, **43**, 46
**Collins, Michael**, member of first Apollo team to land on the Moon, although Collins himself did not land. 58
**Colorado River** 29
**Comet** A cloud of *particles* in orbit around the Sun. 77, **77**
**Concrete** 40, 47
**Condensation** When a gas or vapor turns to liquid.
**Conglomerate**, 40
**Conservation** 47
**Constellations** 84, **84–85**
**Continents** are the largest land masses on Earth. 18–19, 26
**Continental drift** is the process that has moved the continents around. **18–19**, 21
**Continental shelf** 16–17
**Coon Butte**, crater in Australia **76**
**Copernicus**, crater on the Moon 60
**Copernicus, Nicolaus**, early astronomer 49
**Copper** 40, 47
**Core** The center of a planet below the *mantle*. **12–13**, 19; Moon's 60
**Corona** 63
**Cosmic ray** 15, 53
**Cosmonaut** (Soviet astronaut) 55
**Cotopaxi**, tallest active volcano 25
**Crab pulsar** 86
**Crater** 57, 68, 69, **76**; on Mars **73**; on Mercury **68**; on Moon **61**
**Cretaceous period** 18, 32, **32**, 33
**Crust** A planet's hard surface. **12–13**, **18–19**, 20, 25, 30; Moon's crust 60

88

▲ *Cyclones can kill people and damage property. This is Darwin after Cyclone Tracy struck it in 1974.*

**Currents** are warm or cool streams in a liquid or gas. Currents flow through the oceans and the atmosphere. **16–17**, 19, 28
**Cyclones** are areas of low air pressure. They bring stormy weather. Depressions, hurricanes and tornadoes are types of cyclone. See also *anticyclone*.
**Cygnus**, the Swan (constellation of stars) **82**

## D

**Day** 48, 70; on Mars 72
**Deciduous tree** 34, 35, **35**
**Deimos** 72
**Delta** This is a fan-shaped rivermouth. It consists of mud pierced by a network of channels. 28
**Density** A dense substance is one with close-packed *particles*. 67
**Depression** 71. See also *cyclone*.
**Desert** 27, **34–35**; on Mars 72, **73**
**Devonian period** 32, **32**
**Dinosaur** 32
**Dissolve** When some solid substances are mixed with a liquid, the particles vanish. They are said to dissolve. 16, 27
**Dune** 27; on Mars **73**
**Dust** 10, **14**, 21; of comets 77; on Mars **73**; meteoric 77; on Moon 60; and stars 80, 81, 82
**Dwarf star** 79, 80, 81, **80**

## E

**Earth** is the third planet from the Sun and the only one known to have living things. See *Contents page*.
**Earthquake** is a trembling of the Earth's *crust*. 12, 19, **22**, **23**, 25, 60
*Echo* (satellite) 51
**Eclipses** occur when the Earth or Moon cuts off the Sun's light from the other. 63, **63**
**Electricity** 43
**Electro-magnetic waves** are waves of energy such as the Sun gives off. 64
**Electrons** are electrically charged *particles* in *atoms*. 14
**Element** is a substance made up of only one kind of atom. 13, 73
**Elliptical path** is a long looping path, like that of a *comet*.
**Encke's comet** 77, **77**
**Energy** is a measure of the amount of heat a body may produce. 30, **42–43**, 47; of stars 78, 80; of the Sun 64
**Engine** 40, 51, 55
**Eocene epoch** 18, 32
**Epicenter** (of earthquake) 22
**Epoch** (geological) 32
**Equator** is an imaginary circle separating Earth's northern and southern hemispheres. **14**, 17, 35 (See also *hemisphere*.)
**Equinox** This is one of two times each year when the Sun is above the equator and day and night are the same length everywhere. The spring equinox occurs about 21st March and the autumnal equinox about 23rd September.
**Era** (geological) **32**
**Eros** (asteroid) **76**
**Erosion** is the wearing away of the land. 26–29, **26–29** See also *weathering*.
**Escape velocity** is the speed needed to escape the pull of gravity of the Earth or another heavenly body. The greater the pull, the higher the speed must be.
**Estuary** A broad rivermouth is an estuary. 28
**Ethiopian realm** 36, 37
**Evaporation** When a liquid turns to gas, or to vapor, it is said to evaporate. 16, 43
**Evening star.** See *Venus*
**Everest, Mount**, highest mountain in the world 16, 24, 25
**Evergreen** 34, 35, **35**
**Evolution** 33
**Exosphere** 14, **15**
**Expand** means to grow larger 21
*Explorer* (satellites) 51, 52
**Extinct** describes a volcano which is no longer active 21; or a kind of plant or animal that has died out 37

▼ *In a flash of lightning, electricity which has built up in the sky passes to the ground through the Empire State Building.*

## F

**Falling star** 15, 76
**Farming** 30, 38, **39**, 40, 44, 45
**Fault** A crack in the Earth's *crust*. The land on one side may have moved up, down or sideways. 22, **24**, 25
**Feldspar** 27
**Fireball** 10
**Fish** 32, 33, 37, 46
**Flood plain** 28
**Fog** is very low *cloud*.
**Fold** A bend in a layer of rocks.
**Fold mountains** 24–25
**Food** 30, 33, 38, **39**, 40, 44, 45, 46
**Forest** 30, **34–35**, 46
**Fossil** The remains of a dead *organism* from past ages, preserved in the rocks. 25, 33, **33**
**Fossil fuels** 42–43, 46
**Fra Mauro crater** 58
**Friction** is the rubbing of one object against another. 15, 54
*Friendship* (satellite) 55
**Frost** 26, 29
**Full Moon** 62, **62**
**Fuel** is any substance burnt to yield heat or other energy. **42–43**, 47

## G

**Gagarin, Yuri**, first man to orbit Earth 51, **54**, 55
**Galaxy** This is a vast system of *stars*. **78**, **79**, 83, **83**, 86
**Galileo Galilei**, famous astronomer 49
**Gas** is a thin cloud of separate particles which expands to fill the space enclosing them. **8**, 10, 13, 14, 20, 21, **43**, 47, 51, of comets 77; of nebulae 80–81; of planets 75; in the Sun 64, **65**
**Geiger counter** is an instrument which detects and measures radioactivity. 40
*Gemini* (project) 55
**Geophysicists** are scientists who study how the Earth behaves. 12
**Giant planet** 66–67, 75
**Giant star** 49, 81

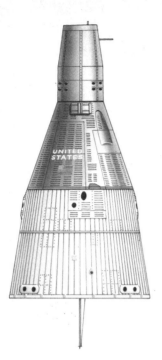

▲ The American spacecraft Gemini, which carried two men.

**Gibbon** 37
**Gibbous** 62, **62**
**Glacier** is a river of ice. 28, **29**
**Glass** 40
**Glenn, John H.**, first American to orbit Earth 55
**Gneiss** 13
**Goddard, Robert**, American scientist 51
**Gold** 46, **47**
**Gold, Thomas**, a British physicist 10
**Grand Canyon** 29
**Granite** 12, **20**, 27, **40**
**Graphite** 40
**Grass** 34, 35
**Grassland** **30**, 34
**Gravel** 28, 40
**Gravimeter** 40
**Gravitation, law of** 49
**Gravity** is a force that tends to pull one object toward another. It is also called gravitation. 10, 13, 51, 54, 56, 59; of Jupiter 67; of Mercury 67, 68; of the Moon 60, 62, 63; and multiple star systems 82; in a space station 87; of the Sun 63, 64
**Great Dividing Range** 24
**Great Red Spot** 75
**Great Rift Valley** 24, **25**
**Gulf of Suez** 19
**Gulf Stream** 16

## H
**Hadley Rille** 58
**Hail** Ice pellets which fall from clouds 17

**Halley's comet** 77, **77**
**Hedgehog** 37
**Helium** 14, 64; on Saturn 75; and stars 80
**Hemisphere** Half a sphere. The Earth is divided by the equator into a northern hemisphere and a southern hemisphere.
**Herschel, William**, famous astronomer 49
**Hertzsprung, Ejnar**, famous astronomer 49
**Hidalgo** (asteroid) 76
**Himalayas** 24, **25**
**Hipparchus**, early astronomer 49
**Hoba Meteorite** 76
**Holarctic realm** 37
**Horsehead nebula** 49
**Hoyle, Fred** 10
**Hubble, Edwin**, famous American astronomer 10, 49
**Humason Comet** 77
**Hurricane** 71. See also *cyclone*.
**Hyades**, star cluster 82
**Hydro-electricity** 43
**Hydrogen** 12, 14; on Jupiter 75; on Saturn 75; in stars 80, 81; in the Sun 64
**Hydrogen bomb** 21

## I
**Icarus** (asteroid) 76
**Ice** is frozen water. 24, 26, 27, 28, **29**; and comets 77
**Icebergs** are floating islands of ice.
**Ice cap** A layer of ice covering land. 28; on Mars 72, 73
**Igneous rock** is rock formed from *magma* or *lava*. 40, **40**
**Ikeya-Seki comet** 77
**Indian Ocean** 16, **16**
**Industry** 44

▼ An iceberg floating in the Antarctic Ocean. Five-sixths of the iceberg is hidden beneath the surface of the water.

▶ The astronomer, William Hershel

**Ionosphere** 14, **15**
**Ions** 15
**Iron** 12, 40, **41**, **47**, and meteorites 76
**Island** 16, 17, **26**

## J
**Jupiter** 66, **66**, 67, 74, 75, **75**
**Jurassic period** 18, 32, **32**

## K
**Kaolin** 27
**Karst** 26
**Kepler**, crater on the Moon 6, **61**
**Kilauea, Mount** 21
**Kohoutek**, comet 77, **77**
**Komarov, Vladimir** First astronaut to die in flight. 55
**Krakatoa** 21

## L
**Labrador Current** 16
**Lake** **29**, 33, 47
**Landslide** 22
**Language** **38**
**Large Magellanic Cloud** 83
**Latitude** is position north or south of the *equator*. It is measured in degrees and shown on maps by lines of latitude parallel with the equator.
**Lava** is *magma* that has escaped on to the Earth's surface from a *volcano*. 20, **21**, 60

**Lead** 46, **47**
**Lemaître, Georges** 10
**Leonids**, meteor shower 76
**Lichen** 34
**Light-year** 78
**Limestone** 25, **26**, 27, 33, 40, **40**
**Longitude** is position east or west on Earth. It is measured in degrees. Each degree is shown on a globe as a line or meridian, between the North Pole and the South Pole. See also *prime meridian*.
**Luna** (spacecraft) 51, 53, 59
**Lunar highlands** 60, **61**
**Lunar module** 59
**Lunar orbiter** (spacecraft) 59
**Lunar probe** 60
**Lyra** 80

## M
**Machines** 40, 44
**Magma** is molten rock. If it cools and hardens it forms *igneous rock*. 19, **20**, 21.
**Magnetometer** 40
**Magnitude** is a measure of a star's brightness.
**Mammal** 30, **30**, **32**, 33, 37
**Man** See *people*.
**Mantle** The layer below a planet's *crust*. **12–13**, **18–19**, 19, 20, 22; Moon's 60
**Marble** 40, **40**
**Marianas Trench** 17
**Mariner** (space probe) 68, 69, 73
**Mars** 66, **66**, 67, 72–73, **72–73**
**Mass** is the amount of *matter* in an object. 64
**Matter** is the atomic material out of which all objects are made. 10
**Mauna Loa** 21, 25
**Meander** A bend in a river. 28, **29**
**Mechanical weathering** 27
**Mercury** (planet) 66, **66**, 67, 68, **68**

*Mercury* (spacecraft) 54, 55
**Mesosphere** 14
**Mesozoic era** 32, **32**
**Metal** 40, **47**, 69
**Metamorphic rock** is *igneous rock* or *sedimentary rock* changed by heat or pressure in the Earth. 20, 40, **40**
**Meteor** 15, 76
**Meteorite** 76–77, **76**
**Meteoroid** This is any small stony or metallic object moving in space. A meteoroid becomes a meteorite when it hits the surface of another object. 54, 60, 68, 70
**Meteorologists** are scientists who study weather. 53
**Micrometeorite** 76
**Mid-Atlantic ridge** 16–17, 25
**Milky Way** 78, **78–79**, 81, 83
**Minerals** are useful substances found in the ground. 40, 46, **47**; on the Moon 62
**Mining** 40, **41**; on the Moon 62
**Minor planet.** See *asteroids*.
**Miocene epoch** 32
**Mira Ceti**, supergiant star 81
**Molten** means melted by heat. 12, 13, 19, 40
**Mongoloid people** 38
**Monkeys** 31, 37
**Month** 56
**Moon**, the 7, 48, 55, 56–63, **56–63**
**Moonquake** 60
**Moons** are objects that orbit *planets*. 10, 64, 67, **74**, **75**; man-made moons 51, 52, 53
**Morning Star** See *Venus*.
**Moss** 34
**Mountain** A high area of land. Its top may be a peak. **18–19**, 20, **24–25**, 26, 28, 30; on Mercury 68; on the Moon 60
**Mud** 28, 33

# N
**NASA** 87
**Natural resources** are useful plants, animals and substances and forms of energy supplied by nature. 46, 47
**Nearctic realm** 36, 37
**Neutron** A type of particle found in the core of an atom. 87
**Nebula** A cloud of gas and particles in outer *space*. 80, 81
**Negroes** 38

▶ *The Trifid Nebula*

**Neotropical realm** 36, 37
**Neptune** 49, 66, 67, **67**, 74, 75, **75**
**Neutron star** 86, 87
**New Moon** 62
**New Stone Age** 45
**Newton, Isaac**, famous astronomer, 48, 49
**Nickel** 12
**Night** 48, 67, 70
**Nile, River**, longest river in the world 28
*Nimbus* (satellite) 53
**Nitrogen** 14, **14**
**Noctilucent cloud** 15
**Northern hemisphere** 70–71
**North Pole** 14, 70
**Nuclear bomb** 47
**Nuclear energy** 43
**Nuclear reaction** happens when a change affects the nucleus (the inner part) of an atom. 64, 80
**Nuclear reactor** 43

# O
**Obsidian** 40
**Oceans** are the largest areas of salt water on Earth. 11, 12, **16–17**, 33, 62
**Ocean floor** 13, **16–17**, 19, 25
**Ocean of Storms** 58
**Ocean trench** 17, 22
**Oil** 42–43, **43**, 46, **47**
**Old Stone Age** 45
**Oligocene epoch** 32
**Orbit** An orbit is the path made by one object moving around another. 54, **58–59**, **66–67**, 70, **76**, 82
**Orbiting astronomical observatory** 53
**Ordovician period** 32
**Ores** are *minerals* that contain metal. 40, **41**
**Organism** A living thing. 30, 42, 43
**Oriental realm** 36, 37, **37**
**Orion nebula** 81, **84**
**Overthrust** 24
**Oxygen** 12, 14, **14**, 51, 68
**Ozone** 14, **15**

# P
**Pacific Ocean** 16, **16**, 22
**Pacific plate** 25
**Palaearctic realm** 36, 37, **37**
**Paleocene epoch** 32
**Paleozoic era** 32
**Pampas** 34
**Pangaea** 19
**Paraffin** 43
**Particles** are tiny pieces of *matter*. 10, 28, 85
**Peak** 24, 25, **29**
**Peat** 43
**Pebble** 27
**Penumbra** 63
**People** 32, 38, 40, **41**, **42–43**, 44–47, **44**, **45**
**Period** (geological) 32
**Permeable rock** is rock that lets water through. 28, 43

**Permian period** 32, **32**
**Perseids**, meteor shower 76
**Phases** (of the Moon) 62, **62**
**Phobos** 72
**Photosphere** 63
*Pioneer* (spacecraft) 51, 53

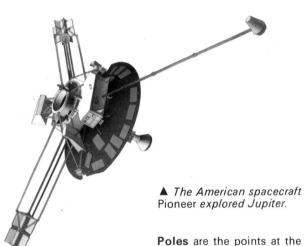

▲ *The American spacecraft* Pioneer *explored Jupiter.*

**Plain** 26, 28, 29, 60
**Planets** are large objects that orbit stars. The Earth is a planet. 10, 48, 64, **66**, 75, 87
**Plants** 30, **30**, 33, **34–35**, 36, 38, 43, 46, 47
**Plate, crustal** The Earth's *crust* consists of giant 'jigsaw' pieces known as crustal plates. **18–19**, 20, **21**, 22, 23, 25
**Platypus** 37
**Pleiades** 82, **82**
**Pleistocene epoch** 32
**Pliocene epoch** 32
**Pluto** 66, 67, **67**, 74, 75, **75**
**Polar regions** are the Arctic and Antarctic. During the winter the Sun remains below the horizon for up to several months. 70, **71**
**Poles** are the points at the ends of the Earth's *axis*. 14; on Mars 72
**Pollen** 30
**Pollution** is anything that makes a place unclean, or unfit for living things. 46, 47
**Population** 45, **45**, 47
**Prairies** 34
**Prehistoric** means before written history. 40
**Pressure** is the force with which one substance presses on another.
**Prime meridian.** This line of *longitude* runs through Greenwich, England. Longitude positions are measured in degrees east or west of it.
*Project Score* (spacecraft) 51
**Propellant** 51
**Prospector** 40, 42
**Pulsar** Regular pulses of *radio waves* come from small stars called pulsars. 86, **86**

# Q
**Quasar** is the most intense kind of energy source known in the universe. 86, **86**
**Quaternary period** 32, **32**
**Quarrying** 40, **41**

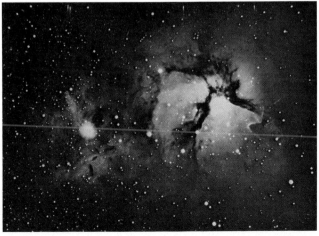

*▲ Radio telescope at Parkes Observatory in New South Wales, Australia.*

## R
**Race** 38
**Radiation** is energy given off as waves, rays or *particles*. 52, 87
**Radioactive** describes a substance whose atoms give off certain rays or *particles*. 40
**Radio telescope** detects *radio waves* received from space. 86
**Radio waves** are types of *electro-magnetic wave*. **15**, 86
**Rain** 11, 13, 14, **17**, 26, 27, 28, 35, 43
**Rainwater** 27
*Ranger* (spacecraft) 59
**Rapids** 28
**Red giant** 78, **80–81**
**Red planet**. See *Mars*.
**Reptile** 32
**Richter Scale** measures the strength of earthquakes. It goes from 0 to over 8. 22
**Rift valley 24**, 25, **25**
**Ring nebula** 80
**River** 13, 16, **17**, **28–29**, 43, 47
**Rock** 12, 13, 19, **20–21**, 22–23, 24–25, 26–27, 28, 33, **40–41**, **42–43**, 46, 56; Moon 59, 60
**Rocket** 50–51, **50–51**, 54, 55, 56
**Rocky Mountains** 24, 25

## S
**Sagittarius**, the Archer (constellation of stars) **83**
**Sahara Desert** 28
**Salts** 16
**San Andreas fault** 23
**San Francisco** 22, 23
**Sand** 27, 28, **28**, 33, 40; on Mars 73
**Sandstone** 33, 40
**Satellites** 52–53, **52–53**
**Saturn** (planet) **66**, 67, **67**, 74, **74**, 75
*Saturn* (rocket) **50**, **51**, 55
**Savanna** 35
**Schiaparelli, Giovanni** 72
**Schmitt, Harrison H.**, member of *Apollo 17* mission 58, **59**
**Scree** 26
**Scorpio**, the Scorpion (constellation of stars) **82**
**Sea bed** 17, 22, 25, 28
**Sea of Serenity** 58, **61**
**Sea of Tranquillity** 58, **61**
**Seas** are areas of salt water smaller only than *oceans*. 13, **16–17**, 26, **26**, 27, 28, 30, 43, 47; 'seas' on Mars 72; on the Moon 60, **61**
**Seasons** are times of year with special kinds of weather. **70**; on Mars 72
**Seawater** 16
**Seaweed** 30, **32**, 33
**Sedimentary rock** is made of sediments, usually laid down in water. 28, 33, **40**
**Sediments** include settled matter like mud and sand. 28, 40
**Seismograph 22**, 40
**Seismometer** 60
**Seven Sisters** 82
**Sewage** 47
**Shale** 33, 40
**Shrub** 34
**Sheep 37**, 46
**Shepard, Alan**, first US spaceman 55
**Shooting star**. See *meteor*.
**Silicate** 13
**Silicon** 12
**Silt** 28
**Silurian period 32**
**Silver 47**
**Sinai Peninsula 19**
**Sky** 14, 48
*Skylab* An American space station which orbited Earth and was manned by three different crews during 1973–74. **55**
**Slate** 40
**Sleet** A mixture of snow and rain. 14, 17
**Smelting** is melting *ores* to extract metals. 40, **41**
**Snow** 14, 17, 28, 29, 34
**Soft landing** A gentle landing rather than a crash.
**Soil** is the loose matter on the surface of the Earth's *crust*. 12, 28, 30, 34, 35, 40, 46; Moon soil 60
**Solar cell** 53
**Solar energy 47**
**Solar prominence 64**
**Solar ray** A ray of energy from the Sun. **15**
**Solar system** A star and the planets and moons that orbit it. 64–67, **64–77**
**Solar wind** 61
**Solstice** is the time of year when the Sun is at its northernmost or its southernmost point.

*▼ The emblem of the American Skylab mission*

**Southern hemisphere** 70–71
**Southern Ocean** 16
**South Pole 14**, 70
*Soyuz* (spacecraft) 55
**Space** is the emptiness beyond the Earth. 7, 10, 48, 51, 53, 54, 78, 87
**Spacecraft** 48, **50**, 52, 53, **55**, **58**, 59
**Spaceman** 54
**Space probe** 51, 52, 53, 59, 68, **72**, 73
**Spaceship** 87
**Space station** 87
**Space suit** 55
**Spectroscope** 82
**Speed of light** 73, 86
**Spore** 30
**Spring** 28, 70
**Springtail** 30
*Sputnik* (satellite) 51, 52, **52**
**Stars** are vast balls of glowing gases. The *Sun* is a star. 7, 8, 10, 14, 48, 64, 66, 78–87, **78–87**
**Star cluster** 82
**Star group** 10, **82–83**
**Star system** 82, **82**
**Steady State theory** 10
**Steam** 11, 13, 43
**Stellar remnant** 81
**Steppe** 34
**Stone** 27, 28, 40; Moon 60
**Storm** 27
**Stratosphere** 14, **15**
**Stream** 28
**Sturgeon** 37
**Suez, Gulf of** 19
**Summer** 34, 70
**Sun** 7, 8, **10**, 14, 48, 64, **64–65**, **66**, 67, 70, **70**, 78, 80–81
**Sunlight** 30, 38, 43, 56, 62
**Sunshine** 33, 53

▲ *An example of a syncline in North Wales*

**Sunspot** is an area on the surface of the Sun which is much cooler than its surroundings and appears darker. **64**
**Supernovae** are old stars that explode and disintegrate. 81, 86
*Surveyor* (spacecraft) 59
**Swamp** 46
**Syncline** A downfold in layers of rocks. 24

## T
**Talcum powder** 40
**Taurus** 82
**Technology** describes the tools and methods used for man's well-being. 40
**Telescope** 48, 60, 78
**Temperate regions** have a mild climate. 34
**Temperature** is hotness measured in degrees. 26, 40, 54; of the Sun 64
**Terrestrial planets** 66–67
**Tertiary period** 32, **32**
**Thawing** 26
**Thíra**, the greatest volcanic explosion 21
**Tidal theory** 10
**Tide** 17, 48, 62–63, **63**
**Time** at any place on Earth depends upon its *longitude*.
**Tin** 46
*Tiros* (satellite) 51
**Tools** 38, 40
**Tracking station** 52–53
**Trade** 45
*Transit* (satellite) 51
**Triassic period** 18, 32, **32**
**Tree** 32, 34–35, **34–35**
**Triple-star system** 82
**Triton** 75
**Tropical** 34, **34**
**Tropics** are regions of the Earth where the Sun is directly overhead at least once a year. 31, **34–35**, 55, 70
**Troposphere** 14, 15

**Tsunami** 22, **23**
**Tundra** 34
**Turbine** 43

## U
**UFO** (unidentified flying object) 87, **87**
**Ultraviolet radiation** consists of *electro-magnetic waves* of short wavelength. 14, 15
**Umbra** 63
**Universe** is all the *matter, energy* and *space* there is. 7, 8, 10, 78, 83, 87
**Uranium** 43, **47**
**Uranus** 49, 66, **67**, 67, 74, **74**, 75
**Ursa Major**, the Great Bear (constellation of stars) 82

## V
**V2 rocket** 51
**Valley** 24, 26, 27, 28, **29**; on Mars 73
**Van Allen Belts** 51, 52
*Vanguard* (satellites) 51, 52, 53
**Variable stars** are stars that vary or seem to vary in brightness. 81
*Venera* capsule 69
**Vesuvius, Mount** 20
**Venus** 66, **66**, 67, 68, 69, **69**, 77
**Verne, Jules** 56
**Viking** (spacecraft) **72**, 73, **73**
**Virgo**, the Virgin (constellation of stars) **83**
**Volcanic bomb** 20
**Volcano** A hole or crack where *magma* escapes from the *crust* 9, 13, 20, 21, **20**, 21, 22, 25, 26; dormant and extinct 20; on the Moon 60
**Volume** is the amount of space a substance takes up.
**Vosges Mountains** 25
*Voskhod* (spacecraft) 55
*Vostok* (spacecraft) 51, **54**, 55
**Vulcano** (island) 20

## W
**Water** 14, **16–17**, 26, 27, 28, 30, 43, **43**, 46, 47; on Mars 73
**Water cycle** The movement of water from the oceans into the air and back. 17
**Waterfall** 28, **42**
**Water table** 28
**Water vapor** 13, 14, 16, 17
**Waterwheel** 43
**Wave** 17, 22, 27, 47
**Wealth** 38, 44, 47
**Weather** 14, **53**
**Weathering** is the breaking up of rock by weather. 26, 27, **26**, **27**, 28
**Wegener, Alfred** 19

**White, Edward** First man to walk in space **54**
**White dwarf** 78, 81
**White people** 38
**Wind** is a moving mass of air. In places winds blow most of the time. The Trade Winds and the Westerlies are such winds. 14, 17, 26, 27, 34, 47
**Winter** 34, 70
**Wood** 43, **43**

## Y
**Year** 67

## Z
**Zebras** 36
**Zinc** 46
**Zoologists** are scientists who study animals. 36

---

### ACKNOWLEDGEMENTS

**Picture Research:** Jackie Cookson
**Photographs:** Endpapers California Institute of Technology; Page 4/5 ZEFA; 6 ZEFA *left*, NASA *top right*, ZEFA *bottom right*; 9 Geoscience Features/Basil Booth; 13 Grønlands Geologiske Undersøgelse/D. Bridgewater; 19 NASA; 21 ZEFA; 22 United States Information Service; 22/23 Associated Press; 24 ZEFA; 25 Pat Morris *top*, Alan Hutchison *bottom*; 26 ZEFA *top*, Michael Chinery *center*, Australian News and Information Bureau *bottom*; 27 ZEFA *left*, Sarah Tyzack *right*; 28 Heather Angel; 29 United States Information Service *top left*, Swiss National Tourist Office *top right*; 31 ZEFA; 33 Sternberg Memorial Museum, Kansas *top*, Imitor/Trustees of the British Museum *bottom*; 34 Brian Hawkes *top*, Heather Angel *center*; 34/35 Heather Angel; 35 Heather Angel *center*, Anthony Bannister/NHPA *bottom*; 36 Canadian Government Tourist Bureau *top*, Alan Root/Survival International *bottom left*, South Africa Tourist Office *bottom right*; 37 Pat Morris *top, center*, Heather Angel, *bottom*; 39 ZEFA; 41 British Steel Corporation *top, bottom left*, Japan Information Service *bottom right*; 42 National Coal Board *right*; 43 British Gas Company; 44 Chicago and North Western Transportation Co *top*; 44/45 ZEFA; 46 Pat Morris *left*, Cohen 500 Group *right*; 47 ZEFA *top*, Atomic Energy Authority *bottom*, 48 Radio Times Hulton Picture Library *top*, Yerkes Observatory *bottom*; 49 California Institute of Technology; 50 NASA; 52 Novosti; 53 NASA; 54 Novosti *top*, NASA *bottom*; 55, 56/57, 58, 59, 60, 61 NASA; 62 California Institute of Technology; 68 NASA; 69 California Institute of Technology *top*, NASA *bottom*; 70 Sonia Halliday; 71 NASA *top*, Meteorological Office *bottom*; 72 California Institute of Technology; 73 NASA *top, bottom*; 74 California Institute of Technology *top, bottom*; 75 California Institute of Technology *top, center, bottom*; 76 American Meteorite Laboratory *top*, California Institute of Technology *bottom*; 77 California Institute of Technology *top, center*, United States Naval Observatory *bottom*; 79 United States Naval Observatory; 80 California Institute of Technology; 81 California Institute of Technology; 82 California Institute of Technology; 83 California Institute of Technology *all*; 86 California Institute of Technology *top*, Lick Observatory *center, bottom*; 86/87 NASA; 87 ZEFA; 88 Shell Limited; 89 Australian Information Service *top*, The Empire State Building *bottom*; 90 National Maritime Museum, London *bottom left*, Trans-Antarctic Expedition *bottom right*; 91 California Institute of Technology; 92 CSIRO *top*, NASA *bottom*; 93 Institute of Geological Sciences, London. Cover: NASA *top left, back*, ZEFA *top right, bottom left, bottom right*.
Cover design by The Tudor Art Agency.

**SALT LAKE COUNTY LIBRARY SYSTEM**
RIVERTON BRANCH

```
J550
Lam        Lambert, D.
             Earth and space
c1979                    7.90
```

PLEASE BRING ME BACK ON TIME
WITH THE DATE DUE CARD IN MY
POCKET.*

*Sorry, I must charge for all lost
or damaged cards.

# SALT LAKE COUNTY LIBRARY SYSTEM